STARS
AND ATOMS

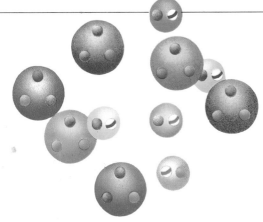

STARS
AND ATOMS

*From the Big Bang
to the Solar System*

STUART CLARK

OXFORD UNIVERSITY PRESS

New York 1995

Contents

1 A Universe of Rules

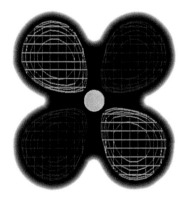

2 The Big Bang

3 Galaxies and Quasars

Project editor   Peter Furtado
Senior editor   John Clark
Editor   Lauren Bourque
Editorial assistant   Marian Dreier

Art editor   Ayala Kingsley
Visualization and artwork   Ted McCausland/ Siena Artworks
Senior designer   Martin Anderson
Designer   Roger Hutchins

Picture manager   Jo Rapley
Picture research   David Pratt
Production   Clive Sparling

Planned and produced by
Andromeda Oxford Ltd
9-15 The Vineyard
Abingdon
Oxfordshire OX14 3PX

© copyright Andromeda Oxford Ltd 1995

Published in the United States of America by
Oxford University Press, Inc.,
200 Madison Avenue
New York, NY 10016

Oxford is a registered trademark of Oxford University Press

Library of Congress Cataloging-in-Publication Data

Clark, Stuart

    Stars and atoms : from the big bang to the solar system / by Stuart Clark

        160 p. 29 x 23cm. -- (The new encyclopedia of science)

    Includes bibliographical reference (p.155) and index

    ISBN 0-19-521087-5: $35.00

    1. Astronomy   2. Cosmology   3. Astrophysics   I. Title   II. Series.

QB43.2.C53   1995

520--dc20   94-30783 CIP

Printing (last digit):9 8 7 6 5 4 3 2 1

Printed in Spain by Graficromo SA, Cordoba

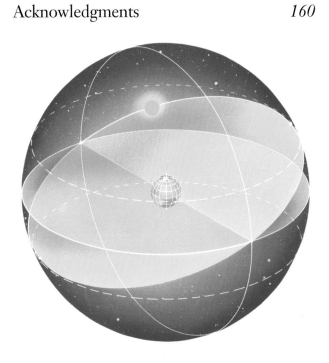

INTRODUCTION

T HE NIGHT SKY, unimaginably deep, is a breathtaking sight. Some three thousand stars can be seen with the naked eye, twinkling points of light that have inspired the human spirit since the dawn of time.

The first attempts to explain the beauty of the heavens took the form of stories of gods and goddesses, heroes and mythical creatures; many ancient cultures used the stars as a focus for their mythological and religious tales. The planets, moving regularly across this background pattern of fixed stars, gave rise to the early science of astrology, which sought to link the fate of humanity with the cyclical motions of the Universe. Assuming the Earth was at the center of the Universe, astrologers observed and recorded these cycles with great care, until discoveries were made that could not be explained within the Earth-centered system.

With the assistance of the first telescopes, astronomy was born as a true science. With ever increasing precision, astronomers observed, plotted and catalogued what they saw. As larger and better telescopes were developed, and other instruments – such as photometers to measure intensity of light, spectrometers to break it down into its various wavelengths, and cameras to record the night sky precisely – the Universe could be studied in exceptional detail. Invisible radiation from the stars, as well as visible light, gave wholly new information, and computers were used to analyze results and provide startling pictures of stars and galaxies. Our knowledge of what the Universe contains has grown hugely in the late 20th century.

In order to make sense of all the new data, astronomers turned for assistance to the physicists who built ever-more powerful machines to conduct experiments for studying the smallest particles of matter. Again, knowledge of this field has grown in recent decades, until most of the fundamental particles and their interactions are reasonably well understood; theoretical physicists have proposed a model of the forces that define the basic rules of the Universe. With the particles providing the words, and the forces supplying the grammar, astronomers can now understand the rich, complex language in which the Universe is written.

Study of the stars, based on data collected from visible-light telescopes, radio telescopes, and detectors for other wavelengths can now reveal extraordinary amounts of information: size, temperature, chemical composition, internal structure, distance and rotation rate, among other facts. Looking more deeply into space, astronomers have discovered thousands of other galaxies, vast collections of stars, dust and gas similar to our own Milky Way galaxy but much farther away.

By applying the theoretical laws of physics, astronomers can explain why objects are as they are; they can also predict what else is waiting to be discovered in space. The Big Bang, which astronomers put forward as the origin of the Universe itself, has been described in step-by-step form, and recent observational evidence has demonstrated the power of much of this description. Even so, problems remain. Perhaps the largest is the surmise that nine-tenths of all the matter in the Universe exists in a form that has yet to be observed and, indeed, cannot be detected directly. What the nature of the "dark matter" is remains controversial.

All this achievement is built on the detailed observation of the faint radiation that reaches our tiny planet from the farthest reaches of the Universe.

While some people feel that the onward march of astronomy is supplanting the traditional concept of God as creator and organizer of the Universe and everything within it, others see the precise account of the ever-growing complexity of matter from the simplicity of the initial event as revealing the underlying mystery, drama, beauty and meaning of the Universe.

THIS BOOK aims to make all this information available to the whole family, from students studying for examinations and projects to adults wanting to bring their scientific knowledge up to date. To achieve this, the book is organized in such a way as to provide readers with a quick answer to a specific query, or to allow them to follow a more detailed account of a particular topic.

At the heart of the book is a 96-page thematic section, made up of 48 major narrative topics, each one richly illustrated to tell the story of a central theme of the book. The strong graphic presentation and the style of writing are designed to make this section the ideal point of departure for the less well-informed reader. Sets of keywords highlighted on each topic spread point the reader to the second major section of the book, a 32-page alphabetic mini-encyclopedia of the subject, containing some 400 entries. This Keyword section can be used entirely independently of the thematic spreads, but this section, too, leads the reader back to thematic topics for a more general account of fundamental subjects.

No region of modern science can be neatly detached from other fields. Astronomy merges into physics and chemistry on one side, planetary and terrestrial geology on another and mathematics (and even philosophy) in other respects. The Knowledge Map, immediately following this Introduction, maps out the entire field of modern science, shows how each area of science is related to another, and defines the major fields. This is followed by a brief Timechart, tracing the development of the subject through the great discoveries since the earliest times.

Finally, to ensure that the volume is of genuine value for reference as well as browsing, the Factfile provides a wealth of hard data, star charts and sky maps, tables and statistics.

KNOWLEDGE MAP
Key Fields of Modern Science

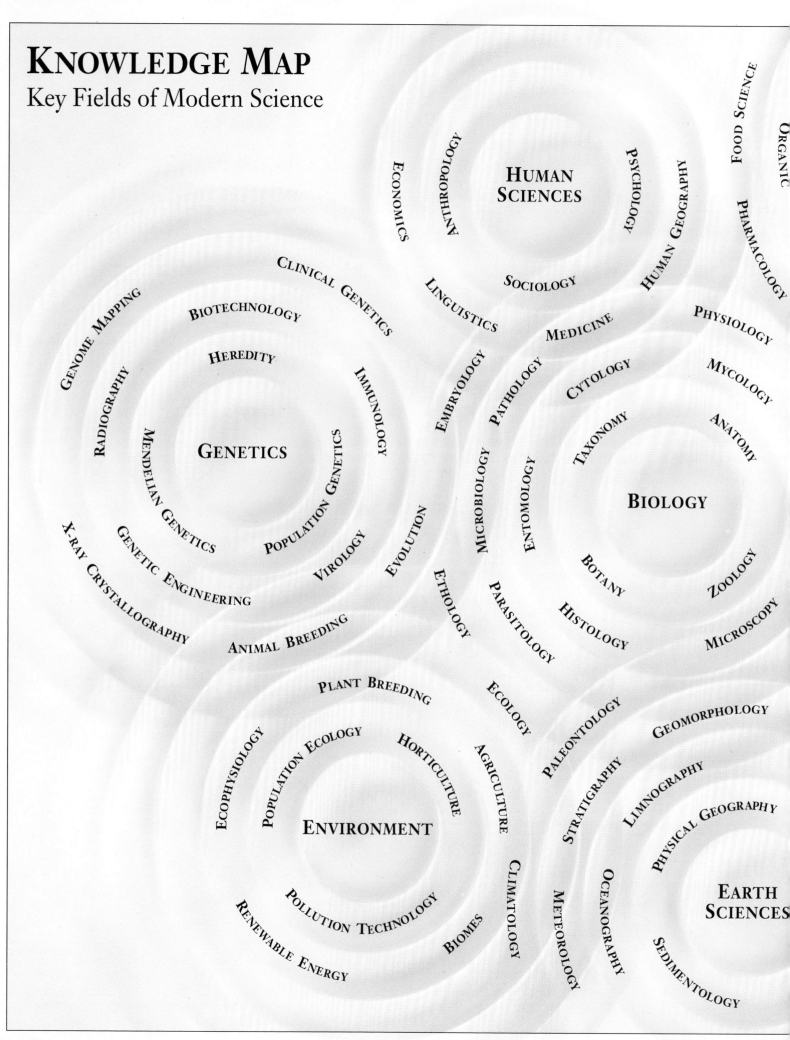

HUMAN SCIENCES

ECONOMICS
ANTHROPOLOGY
PSYCHOLOGY
HUMAN GEOGRAPHY
FOOD SCIENCE
ORGANIC
PHARMACOLOGY

LINGUISTICS
SOCIOLOGY
MEDICINE
PHYSIOLOGY

CLINICAL GENETICS
BIOTECHNOLOGY
GENOME MAPPING
HEREDITY
IMMUNOLOGY
EMBRYOLOGY
PATHOLOGY
CYTOLOGY
MYCOLOGY
ANATOMY

RADIOGRAPHY
GENETICS
TAXONOMY
BIOLOGY

MENDELIAN GENETICS
POPULATION GENETICS
MICROBIOLOGY
ENTOMOLOGY

X-RAY CRYSTALLOGRAPHY
GENETIC ENGINEERING
VIROLOGY
EVOLUTION
BOTANY
ZOOLOGY

HISTOLOGY
MICROSCOPY

ANIMAL BREEDING
ETHOLOGY
PARASITOLOGY

PLANT BREEDING
ECOLOGY
PALEONTOLOGY
GEOMORPHOLOGY

ECOPHYSIOLOGY
POPULATION ECOLOGY
HORTICULTURE
AGRICULTURE
STRATIGRAPHY
LIMNOGRAPHY
PHYSICAL GEOGRAPHY

ENVIRONMENT
CLIMATOLOGY
METEOROLOGY
OCEANOGRAPHY
EARTH SCIENCES

POLLUTION TECHNOLOGY
RENEWABLE ENERGY
BIOMES
SEDIMENTOLOGY

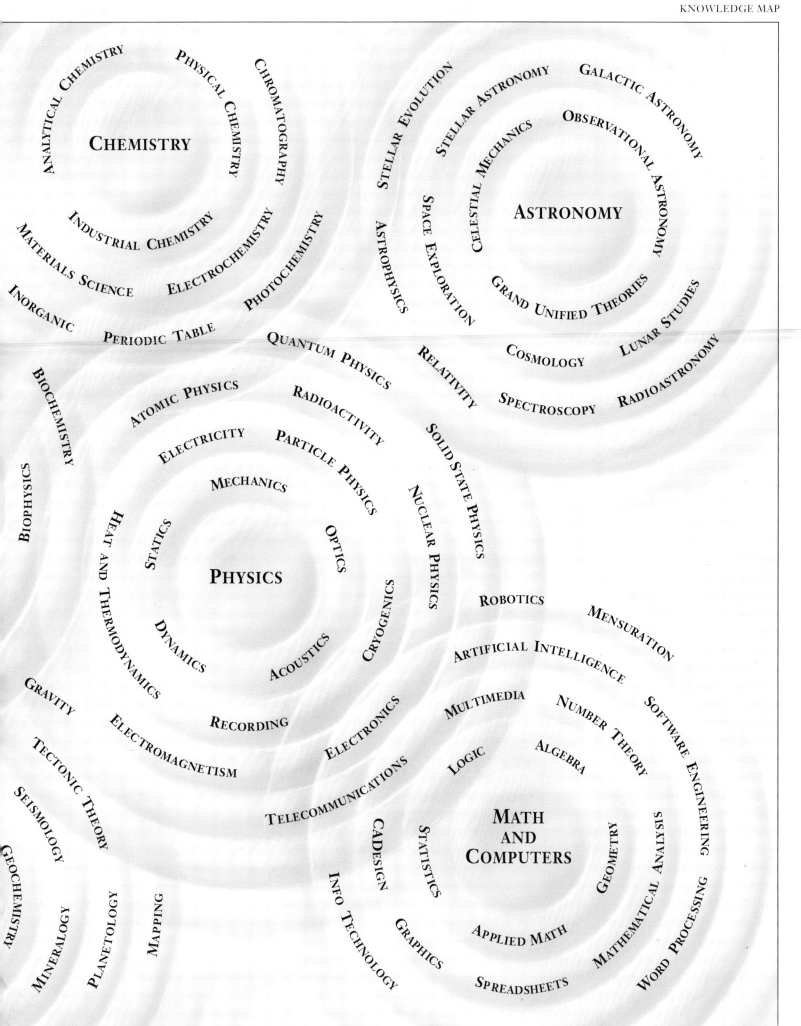

KNOWLEDGE MAP
Modern Astronomy

COSMOCHEMISTRY

The study of the chemisty of celestial objects and interstellar space. It includes identifying elements and compounds, their origins and interactions, such as the energy-generating processes in stars. Cosmochemistry has also revealed the existence of unusual ions and other strange chemicals in space.

ASTROPHYSICS

The study of stars and other stellar objects. Astrophysics covers everything to do with stars, from specialized aspects of STELLAR EVOLUTION to the way in which stars interact with their surroundings. Study of the Sun is classed on its own.

RELATIVITY

A pair of connected theories that describe the properties of light moving through space and the interactions of matter with the spacetime continuum. The special theory of relativity describes light's passage through space and what happens to objects traveling close to the speed of light. The general theory includes the physical phenomenon of gravity.

QUANTUM PHYSICS

The theory that radiant energy is emitted and absorbed in discrete "packets" of energy known as quanta. The quantum of electromagnetic radiation is known as the photon. The way in which quantized energy interacts with matter is known as quantum mechanics. Quantum physics states matter can be waves and waves can be matter.

SPECTROSCOPY

The study of radiation using a prism or diffraction grating to split it into its constituent wavelengths. Absorbed or emitted energy are used to determine which elements are interacting with the radiation. Continuum spectra can provide information about the temperature of objects. Motion can also be deduced from spectra.

RADIO ASTRONOMY

The study of the radio region of the electromagnetic spectrum. The radio band of wavelengths is far larger than the visible region of the spectrum. Certain radio bands have been set aside for the specific use of radio astronomers. The different wavelengths of radiation from an object in space indicate which elements are present in that object. Radio interferometers use two or more radio receivers to provide very precise images.

ASTRONOMICAL PHOTOGRAPHY

The production of visual images of celestical objects. Originally, images were obtained by mounting a camera at the eyepiece of an optical telescope. Today, charge coupled devices (as used in some video cameras) are often used to build up an image electronically. Computer techniques may also be employed to enhance the image, by removing unwanted data and emphasizing key information.

STELLAR ASTRONOMY

The study of stars and their motions through space. This is also known as astrometry. Some nearby stars show parallax and stellar astronomers measure this very precisely so that the entire distance scale of the Universe can be calibrated.

SPACE EXPLORATION

The exploration of space from outside the atmosphere of the Earth. This may involve robotic space probes such as Voyager 2, which explored the outer planets of the Solar System, or manned missions, such as the Apollo missions, which landed men on the Moon in the 1960s and 1970s.

LUNAR STUDIES

The study of the Moon. This is now almost entirely the realm of the geologist and the amateur astronomer. The visible side of the Moon was thoroughly mapped by observers at telescopes before space probes took their photographs. The Apollo astronauts explored the Moon in the late 1960s and early 1970s.

CELESTIAL MECHANICS

The study of the way in which celestial bodies move through space. All celestial bodies move in elliptical, parabolic or hyperbolic orbits. Celestial mechanics predicts the motion of bodies through space and how they interact.

ASTROLOGY

An ancient pseudoscience in which human activities and characteristics are explained by the positions of the Sun, Moon and planets at various times. Astrological observations led to the foundation of astronomy.

PARTICLE PHYSICS

The study of elementary particles, which make up all matter in the Universe. Experiments conducted in high-energy conditions such as particle accelerators provide information about the fundamental forces of nature.

GALACTIC ASTRONOMY

The study of the structure, composition and evolution of the various different types of galaxies. It includes orbital studies of galaxies in clusters, or methods of determining distance to galaxies. One branch of galactic astronomy concerns itself with trying to determine what lies in the center of highly luminous active galaxies.

GRAND UNIFIED THEORIES

A number of independent theories that seek to prove that three of the fundamental forces of nature – electromagnetism, the weak nuclear force and the strong nuclear force – are manifestations of the same "superforce". Many grand unified theories have been postulated but none have been demonstrated to be correct.

QUANTUM MECHANICS

A theory of the interactions of quantum energy with matter. Quantum electrodynamics (QED) is the quantum theory of electromagnetism; quantum chromodynamics (QCD) describes strong force interactions.

STELLAR EVOLUTION

The study of the specific ways in which a star changes over the course of its history. These changes are caused by the nuclear fusion processes which are taking place inside the star. Every change in the type of reactions in the star's core results in some external manifestation that can be observed.

COSMOLOGY

The study of the origin and the fate of the Universe as a whole. A fundamentally theoretical science, cosmology deals with the Big Bang and the events which took place shortly after. These shaped the way in which the Universe evolved and determined the way it will eventually end.

OBSERVATIONAL ASTRONOMY

The study of the Universe based upon taking measurements and readings with a telescope. Any kind of object can be observed, any wavelength can be studied and any form of detector can be used in conjunction with a telescope. All theoretical astronomy and cosmology is ultimately based upon and validated by observations of celestial phenomena.

TIMECHART

MORE THAN 4000 years ago Chinese astronomers were making accurate observations of the stars and planets. Around the same time, Babylonians were also measuring and recording the sky. To them the unchanging pinpoints of light in the night sky revealed the ways of the gods and, even today, the division of the sky into constellations recalls the ancient gods and their doings. Orion, named after the mighty hunter of Greek mythology, wheels overhead during the northern winter; Scorpius, the scorpion said to have stung Orion to death, is found on the opposite side of the sky. Cepheus, his wife Cassiopeia, their daughter Andromeda and her savior Perseus give their names to constellations that lie side by side in the sky. But as science develops, the myths behind the names are put aside.

The great restraint on astronomy was the limitation of the human senses. We cannot see the faintest stars, nor distinguish fine details of the night sky – although early astronomers did manage remarkable feats of observation with the naked eye. Instruments were developed to assist. Tycho Brahe, a young

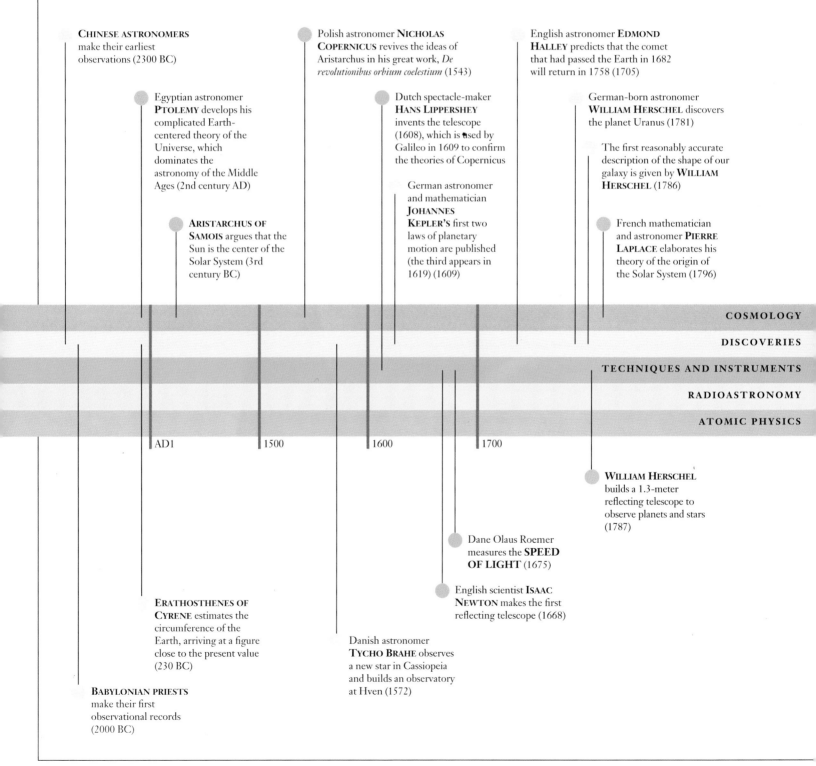

CHINESE ASTRONOMERS make their earliest observations (2300 BC)

Egyptian astronomer **PTOLEMY** develops his complicated Earth-centered theory of the Universe, which dominates the astronomy of the Middle Ages (2nd century AD)

ARISTARCHUS OF SAMOIS argues that the Sun is the center of the Solar System (3rd century BC)

Polish astronomer **NICHOLAS COPERNICUS** revives the ideas of Aristarchus in his great work, *De revolutionibus orbium coelestium* (1543)

Dutch spectacle-maker **HANS LIPPERSHEY** invents the telescope (1608), which is used by Galileo in 1609 to confirm the theories of Copernicus

German astronomer and mathematician **JOHANNES KEPLER'S** first two laws of planetary motion are published (the third appears in 1619) (1609)

English astronomer **EDMOND HALLEY** predicts that the comet that had passed the Earth in 1682 will return in 1758 (1705)

German-born astronomer **WILLIAM HERSCHEL** discovers the planet Uranus (1781)

The first reasonably accurate description of the shape of our galaxy is given by **WILLIAM HERSCHEL** (1786)

French mathematician and astronomer **PIERRE LAPLACE** elaborates his theory of the origin of the Solar System (1796)

COSMOLOGY

DISCOVERIES

TECHNIQUES AND INSTRUMENTS

RADIOASTRONOMY

ATOMIC PHYSICS

AD1 1500 1600 1700

WILLIAM HERSCHEL builds a 1.3-meter reflecting telescope to observe planets and stars (1787)

Dane Olaus Roemer measures the **SPEED OF LIGHT** (1675)

English scientist **ISAAC NEWTON** makes the first reflecting telescope (1668)

ERATHOSTHENES OF CYRENE estimates the circumference of the Earth, arriving at a figure close to the present value (230 BC)

Danish astronomer **TYCHO BRAHE** observes a new star in Cassiopeia and builds an observatory at Hven (1572)

BABYLONIAN PRIESTS make their first observational records (2000 BC)

Danish nobleman, who was astonished by the sudden appearance of a new star in 1572, built enormous instruments to measure star positions accurately. The greatest leap forward occurred in 1609 when the Italian astronomer Galileo first turned his telescope to the sky. Immediately, he saw thousands of stars where before there had been an undefined glow. He saw wonders such as mountains on the Moon, the satellites of Jupiter, and contemporaries simply could not believe their eyes when they looked into the telescope.

Ever since Galileo, each newly invented instrument has carried astronomy into new areas. The limitations of the human eye have been left far behind. The reflecting telescope, first built by English scientist Isaac Newton, circumvented most of the limitations of the lens or refracting telescope, and led on to the large reflectors of the modern age. The 500-centimeter Hale telescope on Mount Palomar, California, United States, is so sensitive that it could detect a candle shining 25,000 kilometers away. At the same time astronomers have developed instruments to detect radiations invisible to the eye. German-born English

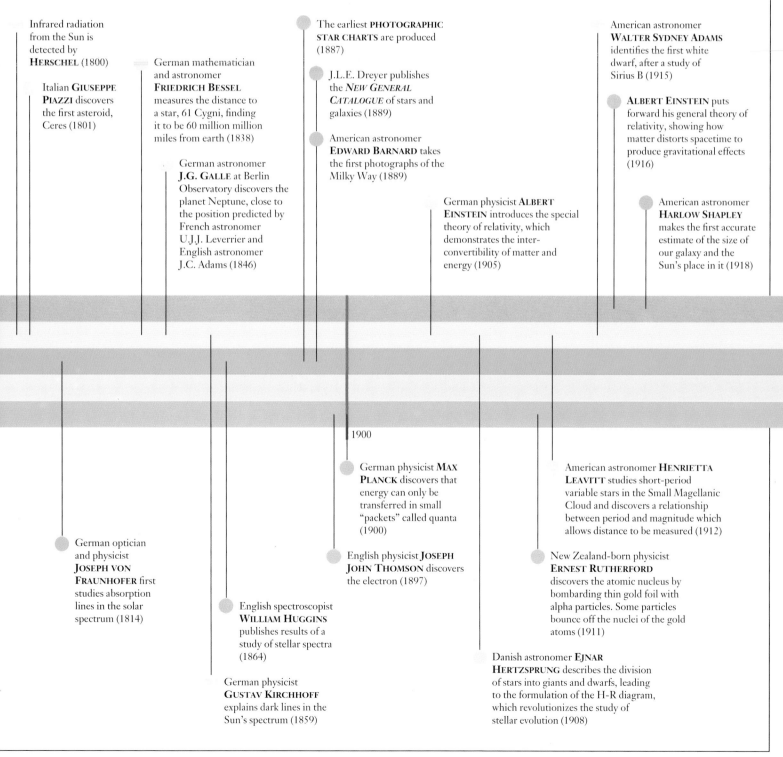

Infrared radiation from the Sun is detected by **HERSCHEL** (1800)

Italian **GIUSEPPE PIAZZI** discovers the first asteroid, Ceres (1801)

German mathematician and astronomer **FRIEDRICH BESSEL** measures the distance to a star, 61 Cygni, finding it to be 60 million million miles from earth (1838)

German astronomer **J.G. GALLE** at Berlin Observatory discovers the planet Neptune, close to the position predicted by French astronomer U.J.J. Leverrier and English astronomer J.C. Adams (1846)

The earliest **PHOTOGRAPHIC STAR CHARTS** are produced (1887)

J.L.E. Dreyer publishes the *NEW GENERAL CATALOGUE* of stars and galaxies (1889)

American astronomer **EDWARD BARNARD** takes the first photographs of the Milky Way (1889)

German physicist **ALBERT EINSTEIN** introduces the special theory of relativity, which demonstrates the inter-convertibility of matter and energy (1905)

American astronomer **WALTER SYDNEY ADAMS** identifies the first white dwarf, after a study of Sirius B (1915)

ALBERT EINSTEIN puts forward his general theory of relativity, showing how matter distorts spacetime to produce gravitational effects (1916)

American astronomer **HARLOW SHAPLEY** makes the first accurate estimate of the size of our galaxy and the Sun's place in it (1918)

1900

German physicist **MAX PLANCK** discovers that energy can only be transferred in small "packets" called quanta (1900)

English physicist **JOSEPH JOHN THOMSON** discovers the electron (1897)

American astronomer **HENRIETTA LEAVITT** studies short-period variable stars in the Small Magellanic Cloud and discovers a relationship between period and magnitude which allows distance to be measured (1912)

New Zealand-born physicist **ERNEST RUTHERFORD** discovers the atomic nucleus by bombarding thin gold foil with alpha particles. Some particles bounce off the nuclei of the gold atoms (1911)

German optician and physicist **JOSEPH VON FRAUNHOFER** first studies absorption lines in the solar spectrum (1814)

English spectroscopist **WILLIAM HUGGINS** publishes results of a study of stellar spectra (1864)

German physicist **GUSTAV KIRCHHOFF** explains dark lines in the Sun's spectrum (1859)

Danish astronomer **EJNAR HERTZSPRUNG** describes the division of stars into giants and dwarfs, leading to the formulation of the H-R diagram, which revolutionizes the study of stellar evolution (1908)

astronomer William Herschel detected infrared radiation in 1800. Today, astronomers view the Universe using infrared, ultraviolet, X-ray, gamma, and radio radiation, as well as visible light. The most developed of these is radioastronomy. Radio telescopes can see details far finer than optical telescopes can. Even so, the optical Space Telescope can see an object the size of a small coin 725 kilometers away – the distance from New York to Detroit.

Astronomers also learned to extract information from the light their telescopes captured. Spectroscopy, the spreading of light

into a spectrum, revealed the chemical composition of the stars and nebulae to Gustav Kirchhoff and William Huggins. Later researchers were able to deduce the temperature and pressures inside stars. Finally, spectroscopy revealed a great secret – that the Universe is expanding.

The key was red shift: the change in a star's spectrum caused by its motion away from the Earth. In 1930, studying the spectra of distant galaxies, the American Edwin Hubble showed they were all moving away from the Earth. This result had been predicted by

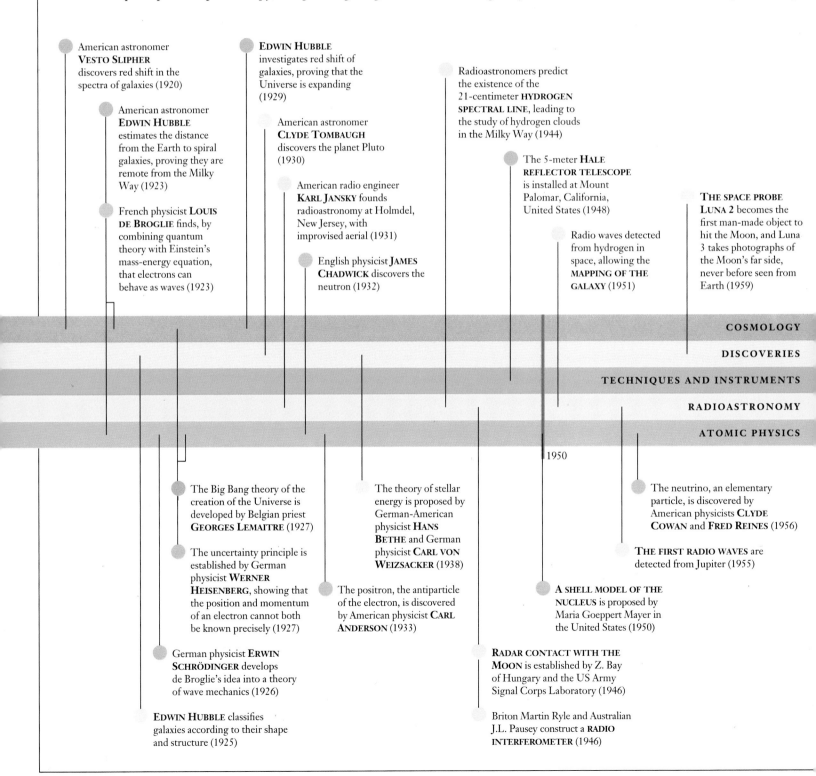

American astronomer **VESTO SLIPHER** discovers red shift in the spectra of galaxies (1920)

American astronomer **EDWIN HUBBLE** estimates the distance from the Earth to spiral galaxies, proving they are remote from the Milky Way (1923)

French physicist **LOUIS DE BROGLIE** finds, by combining quantum theory with Einstein's mass-energy equation, that electrons can behave as waves (1923)

EDWIN HUBBLE investigates red shift of galaxies, proving that the Universe is expanding (1929)

American astronomer **CLYDE TOMBAUGH** discovers the planet Pluto (1930)

American radio engineer **KARL JANSKY** founds radioastronomy at Holmdel, New Jersey, with improvised aerial (1931)

English physicist **JAMES CHADWICK** discovers the neutron (1932)

Radioastronomers predict the existence of the 21-centimeter **HYDROGEN SPECTRAL LINE**, leading to the study of hydrogen clouds in the Milky Way (1944)

The 5-meter **HALE REFLECTOR TELESCOPE** is installed at Mount Palomar, California, United States (1948)

Radio waves detected from hydrogen in space, allowing the **MAPPING OF THE GALAXY** (1951)

THE SPACE PROBE LUNA 2 becomes the first man-made object to hit the Moon, and Luna 3 takes photographs of the Moon's far side, never before seen from Earth (1959)

COSMOLOGY

DISCOVERIES

TECHNIQUES AND INSTRUMENTS

RADIOASTRONOMY

ATOMIC PHYSICS

1950

The Big Bang theory of the creation of the Universe is developed by Belgian priest **GEORGES LEMAITRE** (1927)

The uncertainty principle is established by German physicist **WERNER HEISENBERG**, showing that the position and momentum of an electron cannot both be known precisely (1927)

German physicist **ERWIN SCHRÖDINGER** develops de Broglie's idea into a theory of wave mechanics (1926)

EDWIN HUBBLE classifies galaxies according to their shape and structure (1925)

The theory of stellar energy is proposed by German-American physicist **HANS BETHE** and German physicist **CARL VON WEIZSACKER** (1938)

The positron, the antiparticle of the electron, is discovered by American physicist **CARL ANDERSON** (1933)

A SHELL MODEL OF THE NUCLEUS is proposed by Maria Goeppert Mayer in the United States (1950)

RADAR CONTACT WITH THE MOON is established by Z. Bay of Hungary and the US Army Signal Corps Laboratory (1946)

Briton Martin Ryle and Australian J.L. Pausey construct a **RADIO INTERFEROMETER** (1946)

The neutrino, an elementary particle, is discovered by American physicists **CLYDE COWAN** and **FRED REINES** (1956)

THE FIRST RADIO WAVES are detected from Jupiter (1955)

Georges Lemaître and by Einstein's general theory of relativity. Two pieces of evidence seem to have clinched the case for the expanding Universe, and the "Big Bang" theory of the origin of the Universe which follows from it. In 1965, Americans Arno Penzias and Robert Wilson detected the cosmic microwave background radiation – the last flickers of radiation left from the immensely hot explosion at the beginning of the Universe. In 1992 the satellite COBE detected variations in the background radiation which indicated that the conditions existed for the formation of galaxies only 300,000 years after the Big Bang. This observation completed a consistent picture of the birth of the Universe.

The conditions immediately after the Big Bang continue to fascinate researchers. The Universe was then immensely hot and matter consisted of a soup of fundamental particles, such as quarks. To understand this, scientists turned to quantum theory, the science of subatomic particles. Thus the science of the largest structures in the Universe relies on the theory of the smallest.

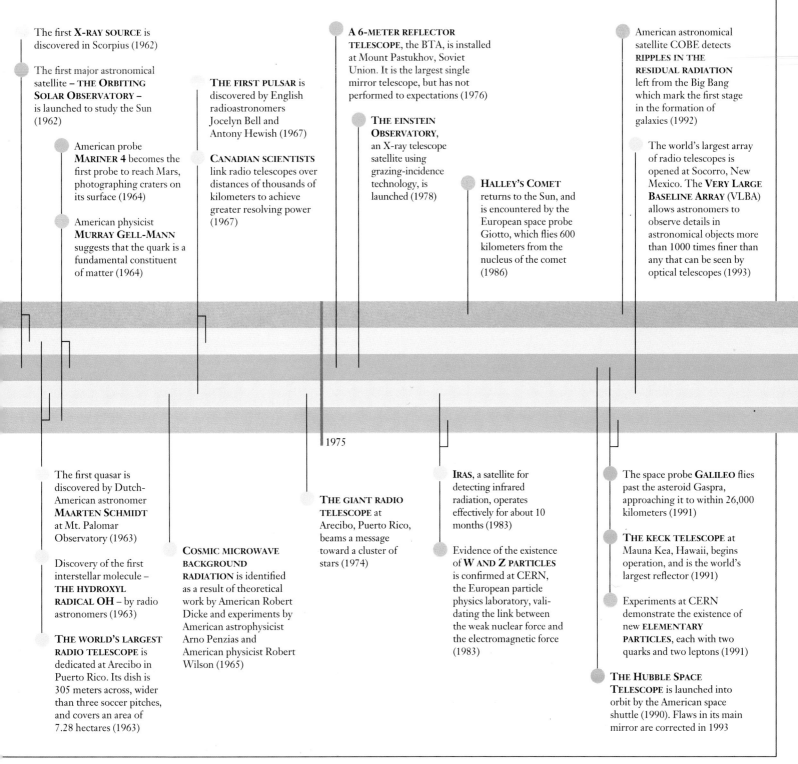

The first **X-RAY SOURCE** is discovered in Scorpius (1962)

The first major astronomical satellite – **THE ORBITING SOLAR OBSERVATORY** – is launched to study the Sun (1962)

American probe **MARINER 4** becomes the first probe to reach Mars, photographing craters on its surface (1964)

American physicist **MURRAY GELL-MANN** suggests that the quark is a fundamental constituent of matter (1964)

THE FIRST PULSAR is discovered by English radioastronomers Jocelyn Bell and Antony Hewish (1967)

CANADIAN SCIENTISTS link radio telescopes over distances of thousands of kilometers to achieve greater resolving power (1967)

A 6-METER REFLECTOR TELESCOPE, the BTA, is installed at Mount Pastukhov, Soviet Union. It is the largest single mirror telescope, but has not performed to expectations (1976)

THE EINSTEIN OBSERVATORY, an X-ray telescope satellite using grazing-incidence technology, is launched (1978)

HALLEY'S COMET returns to the Sun, and is encountered by the European space probe Giotto, which flies 600 kilometers from the nucleus of the comet (1986)

American astronomical satellite COBE detects **RIPPLES IN THE RESIDUAL RADIATION** left from the Big Bang which mark the first stage in the formation of galaxies (1992)

The world's largest array of radio telescopes is opened at Socorro, New Mexico. The **VERY LARGE BASELINE ARRAY** (VLBA) allows astronomers to observe details in astronomical objects more than 1000 times finer than any that can be seen by optical telescopes (1993)

1975

The first quasar is discovered by Dutch-American astronomer **MAARTEN SCHMIDT** at Mt. Palomar Observatory (1963)

Discovery of the first interstellar molecule – **THE HYDROXYL RADICAL OH** – by radio astronomers (1963)

THE WORLD'S LARGEST RADIO TELESCOPE is dedicated at Arecibo in Puerto Rico. Its dish is 305 meters across, wider than three soccer pitches, and covers an area of 7.28 hectares (1963)

COSMIC MICROWAVE BACKGROUND RADIATION is identified as a result of theoretical work by American Robert Dicke and experiments by American astrophysicist Arno Penzias and American physicist Robert Wilson (1965)

THE GIANT RADIO TELESCOPE at Arecibo, Puerto Rico, beams a message toward a cluster of stars (1974)

IRAS, a satellite for detecting infrared radiation, operates effectively for about 10 months (1983)

Evidence of the existence of **W AND Z PARTICLES** is confirmed at CERN, the European particle physics laboratory, validating the link between the weak nuclear force and the electromagnetic force (1983)

The space probe **GALILEO** flies past the asteroid Gaspra, approaching it to within 26,000 kilometers (1991)

THE KECK TELESCOPE at Mauna Kea, Hawaii, begins operation, and is the world's largest reflector (1991)

Experiments at CERN demonstrate the existence of new **ELEMENTARY PARTICLES**, each with two quarks and two leptons (1991)

THE HUBBLE SPACE TELESCOPE is launched into orbit by the American space shuttle (1990). Flaws in its main mirror are corrected in 1993

Astronomy
KEYWORDS

aberration of starlight

The apparent displacement of a star from its real position on the **celestial sphere** caused by the Earth's motion in its orbit around the Sun. Aberration was discovered by James Bradley in the late 1720s, when he was trying to measure parallax. Angular changes of up to 20 arcseconds in the positions of the stars were measured. These are much larger than those expected due to parallax. Alpha Centauri, the nearest star, has the greatest parallax but even that is only 0.76 arcseconds. The measurements were finally explained by two other factors: that light has a finite velocity and the Earth is in constant motion. In the time it takes light to reach our eye, our vantage point has moved and this velocity is superimposed on that of the light from the star. Because our minds extrapolate light sources in a straight line, we appear to see the star in a direction that is slightly displaced from its true position.

absolute magnitude

A measurement of a star's brightness. In effect, it is the **apparent magnitude** of the star if it were placed at 10 parsecs (32.6 light-years) from the Earth. The apparent magnitude of a star depends on its **luminosity** and its distance. If all stars were placed at the same distance then their apparent magnitudes would depend solely on their luminosities. Thus, absolute magnitudes are true indicators of how much light each star emits. The Sun has an absolute magnitude of 4.8, which is relatively faint.

absolute zero

In theory, the lowest temperature it is possible to attain, equivalent to -273° Celsius and zero kelvin. Motion slows as matter cools until, at absolute zero, it stops.

absorption nebula

A region of dust and gas that has no stars to illuminate it from within. It obscures any light from stars or **emission nebulas** that may lie behind it. Absorption nebulas are usually very cold: typical temperatures are 10 to 20 kelvin. They provide the conditions for complex chemistry to take place. This is thought to occur between atoms that adhere to the surfaces of the dust particles and come into contact. Gravitational instabilities arise in these cold clouds, causing regions to collapse and form new stars. *See also* **Barnard objects**, **Bok globules** and **protostars**.

acceleration

The rate at which the velocity of a moving object changes. Velocity, a vector quantity, is speed in a given direction; an object can accelerate by changing its speed or by changing its direction. The units of acceleration, itself a vector quantity, are those of velocity per unit time, eg meters per second, per second (m/sec^2).

accretion disk

A disk-like structure of dust and gas that forms around the massive object at the center of a region undergoing gravitational collapse. It occurs because the regions of collapse are not stationary but move through space. They are in orbit around the center of the galaxy, and so have **angular momentum**. As they fall into the potential well of the massive object, instead of falling straight onto the central body, they spiral downward. This causes the material to build up in a disk around the object. The more massive the object the deeper its potential well and the faster the material in the accretion disk will swirl around it. The more heating the disk undergoes, the more energy it will release in the form of electromagnetic radiation. Accretion disks are, primarily, found around **protostars** and **black holes**.

active galactic nucleus (AGN)

A galactic nucleus in which vast amounts of energy in the form of electromagnetic radiation are being liberated, sometimes enough to outshine all the other stars in the galaxy combined. Most astronomers believe that at the center of an active galactic nucleus lies a **black hole**. A minority believe that AGN can be explained as **starbursts** and **multiple supernovae**. *See also* **active galaxy**.

active galaxy

A galaxy in which an unusual, perhaps violent, phenomenon is releasing energy, usually from the galactic nucleus and unlike the way in which energy is normally released by stars. Active galaxies comprise 10 percent of the population of galaxies that can be viewed by astronomers from Earth. There are many different classifications of active galaxies, such as **Seyfert galaxies**, **quasars**, **radio galaxies** and **blazars**. A unified theory of active galaxies attempts to explain all types of active galaxy as being intrinsically the same kind of object. *See also* **active galactic nucleus**.

> **CONNECTIONS**
>
> SCALE OF THE UNIVERSE 76
> THE MILKY WAY 92
> ACTIVE GALAXIES 96
> ENERGY MACHINE 98

active optics

A system of electronic supports that deform the mirror inside a telescope into the most precise possible shape to maximize its performance. This allows much larger, lighter mirrors to be constructed because they no longer need to support their own weight. *See also* **adaptive optics**.

adaptive optics

A system of fast-reacting electronic supports that deform the mirror inside a telescope to compensate for the way in which the Earth's atmosphere distorts our view of celestial objects. In effect it detects the twinkling of stars and moves the telescope's mirror to cancel this out. This process stops the image from blurring and allows more detail to be seen. *See also* **active optics**.

albedo

The reflectivity of an object. The albedo is expressed as the ratio of the amount of light reflected by an object to the amount of light falling on it.

Alpha Centauri

A triple star system that has the distinction of being the closest star system to the Sun. It exists at a distance of 4.3 **light years**. The components are Alpha Centauri A and B and Proxima Centauri, which is slightly closer than the other two. Alpha Centauri can be seen in the southern sky in the constellation of Crux as a bright star of **apparent magnitude** –0.1. In reality, it is a star very similar to the Sun, because both have **spectral classifications** of G2 and both are on the **main sequence**.

alpha particle

The fast-moving nucleus of a helium atom, comprising two protons and two neutrons. It is emitted in a type of radioactive decay called alpha decay. *See also* **atomic nucleus**.

altazimuth mounting

A method of mounting a telescope using two rotation axes, one parallel to the horizon and the other perpendicular to it. It is the simplest movable mounting for a telescope. Because stars appear to move across the **celestial sphere** due to the rotation of the Earth, telescopes must track them. With an altazimuth mounting, the telescope must be driven about both axes. **Equatorial mountings** are a better design, since they need to be driven around only one axis, but they are much more expensive. *See also* **altitude** and **azimuth**.

altitude

An angular measurement of the vertical distance between the horizon and a celestial object. *See also* **azimuth** and **celestial sphere**.

Andromeda galaxy

The nearest spiral galaxy to the **Milky Way**, and the most distant object visible to the naked eye. It is 2.2 million light-years away and contains 4×10^{11} stars. The Andromeda galaxy is believed to be a twin of the Milky Way. Their combined gravitational influence dominates the **Local Group** of galaxies. It is designated M31 in the **Messier Catalogue** and NGC 224 in the **New General Catalogue**, and is surrounded by satellite galaxies and globular clusters.

> **CONNECTIONS**
>
> SCALE OF THE UNIVERSE 76
> CLUSTERS AND VOIDS 94

angular momentum

The mass of an orbiting body multiplied by its velocity and the radius of its orbit. According to the laws of physics, the angular momentum of any orbiting body remains constant at all points in the orbit. If the orbit is elliptical, the radius is variable. If the radius alters, the velocity changes. This is why planets in elliptical orbits travel faster at **periastron**, when the radius is small, and more slowly at **apastron**, when the radius is large. *See also* **Kepler's laws of planetary motion**.

antimatter

A particle with opposite properties to those of its matter counterpart. The antimatter counterpart of an electron, known as a positron, has a positive charge, equal in strength to the negative charge of the electron. It also has an opposite **spin**. Antimatter is created in **pair production**. When antimatter comes into contact with its matter counterpart, the two particles are instantly annihilated by having their **rest mass** turned into energy and released as electromagnetic radiation. The Universe is predominantly composed of matter, because matter is produced with a slight bias over antimatter.

> **CONNECTIONS**
>
> THE PARTICLE FAMILY 62
> EXCLUSION AND UNCERTAINTY 66
> INFANT UNIVERSE 82
> BLACK HOLES 130

apastron

The point at which a celestial object's elliptical orbit around a star has its maximum radius. *See also* **periastron**.

aphelion

The greatest distance from the Sun reached by an object, such as a planet, in an elliptical orbit around it.

apogee

The greatest distance from the Earth reached by an object, such as the Moon, in an elliptical orbit around it.

apparent magnitude

The brightness of a celestial object as it appears from the Earth. It does not take into account the distance of the object from the Earth, and so is not a reliable indication of any physical property of the object.

Aquarius

The Water Bearer, a large but faint constellation on the **ecliptic**; the 11th sign of the **zodiac**.

arcminute

An angular measurement corresponding to one sixtieth of a degree. *See also* **arcsecond**.

arcsecond

An angular measurement corresponding to one three thousand six hundredth part of a degree: one sixtieth of an **arcminute**.

Aries

The Ram, the first constellation of the zodiac; a winter constellation in northern skies.

asterism

A subset of stars within a constellation which form a separate pattern. The Plow, or Big Dipper, is an asterism within Ursa Major.

asteroid

Any of many rocky objects, most of which orbit the Sun between Mars and Jupiter. Several thousand asteroids are known to exist but there may be 100,000, many of them too small or too faint to be detected from the Earth. The largest asteroid, Ceres, has a diameter of just over 1000 km. The smallest detected asteroids have diameters of approximately 100 m. **Comets**, meteoroids and asteroids make up the minor bodies of the Solar System. They are the leftover planetessimals from the formation of the Solar System. Some asteroids travel in eccentric, elliptical orbits which cross the orbit of the Earth. These are known as Earth-crossing asteroids.

astrometric binary

A star in a binary star system, in which the gravitational influence of its companion causes it to oscillate around a mean position.

astronomical unit (AU)

A unit of astronomical measurement, defined as the mean radius of the Earth's orbit around the Sun. (1 AU = 149,597,870 km)

astronomy

The study of the Universe. It was traditionally concerned with mapping the heavens and understanding how celestial objects moved. Astrophysics, which explores the behavior of things in the Universe, is now more prominent.

astrophysics

The scientific discipline that seeks to explain why celestial objects, and the Universe in general, behave in the way that they do. Its basic assumption is that the laws of physics are the same everywhere in the Universe. *See also* **astronomy**.

atom

The smallest unit that matter can be divided into and still retain its chemical identity. Atoms are complex arrangements of three types of fundamental particles. Protons and neutrons provide the bulk of an atom's mass and are concentrated into a tiny central region known as the atomic nucleus. The nucleus is held together by the **strong nuclear force** which acts between the constituent particles. The protons in the nucleus give it a positive electrical charge. Whereas the radius of an atom is 10^{-10} m, the atomic nucleus is about 10,000 times smaller at 10^{-14} m. The extra volume surrounding the nucleus is due to the presence of the electrons. They are much lighter than the protons and neutrons and carry a negative electrical charge. The electrons are held around the nucleus by **electromagnetism** and exist in electron shells. Within each shell, energy levels allow them to exist in **quantum states** depending upon their energy, in accordance with **Pauli's exclusion principle**. In an atom that has not undergone ionization, the number of electrons equals the number of protons. When ionization occurs, one or more electrons are provided with enough energy to escape the atom altogether. What remains is known as an ion. It is now positively charged and behaves in a different way from the neutral atom. The number of protons in the nucleus of an atom defines which chemical element it is. There are 106 known chemical elements and these can be arranged in a classification sequence known as the Periodic Table. The number of neutrons in a nucleus can also vary. Such variations in the same element are known as its isotopes. Most atoms have a preferred number of neutrons at which their nuclei are stable. Atoms with more or less neutrons than the stable configuration often undego radioactive decay. The simplest atom is the hydrogen atom. It has a nucleus of one proton and a single electron in orbit around it, but no neutron. Deuterium, a stable isotope of hydrogen, does have one neutron in its nucleus. *See also* **atomic absorption**, **atomic emission**, **atomic number** and **molecule**.

atomic absorption

The process by which atoms absorb certain preferred wavelengths of electromagnetic radiation. *See also* **atomic emission**, **energy levels** and **spectral lines**.

atomic emission

The process by which atoms emit certain, preferred wavelengths of electromagnetic radiation. *See also* **atomic absorption**, **energy levels** and **spectral lines**.

atomic mass

The number of protons and neutrons in an atomic nucleus. It is usually denoted by the symbol A. *See also* **atomic number**.

atomic nucleus

The tiny central region of an atom which contains the bulk of its mass. It is held together by the **strong nuclear force** and consists of the fundamental particles protons and neutrons. *See also* **element** and **isotope**.

atomic number

The number of protons in an atomic nucleus. It is usually denoted by the symbol Z. *See also* **atomic mass**.

aurora

The glow that appears in the night sky at high northern and southern latitudes, commonly known as the Northern or Southern Lights. It is produced when charged particles – electrons and protons – collide with atoms and molecules in the Earth's upper atmosphere. Some of the energy of collision is converted into **electromagnetic radiation** in the visible spectrum, causing the sky to glow. **Solar wind** produces a vast amount of electrons and protons, which can cause auroral storms. Usually the aurorae occur within 20° of the north and south magnetic poles but, at times of high solar activity, the storms can migrate farther toward the Equator.

azimuth

The angle a celestial object makes from due north, when measured in an eastward direction along the horizon. When combined with altitude it provides a method for referencing objects on the **celestial sphere**.

barred-spiral galaxy

A form of disk **galaxy** in which the spiral arms are attached to the galactic nucleus by a straight bar of stars. Such galaxies are part of the Hubble classification of galaxies. *See also* **Hubble tuning fork diagram**.

Barnard object

Another name for a dark **nebula** that obscures the light from background stars. *See also* **absorption nebula** and **Bok globule**.

baryon

A particle consisting of three elementary particles known as quarks. Baryons are the particles that make up atomic nuclei. *See also* **hadrons**, **neutrons** and **protons**.

baryon degenerate matter

A form of **degenerate matter** in which electrons are forced into the atomic nucleus by the weight of overlying material. The electrons and protons merge, forming neutrons; these, obeying **Pauli's exclusion principle**, provide degeneracy pressure to halt further collapse. *See also* **electron degenerate matter** and **Oppenheimer–Volkoff limit**.

Becklin–Neugebauer object

The first star to be discovered using infrared detection methods. Deeply embedded within the Orion star-forming nebula, it is invisible at optical wavelengths. This is because the light is completely scattered or absorbed on account of the high density of dusty material in the Orion nebula. Investigation has shown that the Becklin–Neugebauer object has a **spectral classification** of type B and is currently on the **main sequence** of stellar evolution.

Beta Pictoris

A star shown by infrared observations to have a dusty disk of material around it. Once believed to be an accretion disk, from which planets form, it is now known that it is far too large for that to be so. With a diameter 10 times that of Pluto's orbit, the disk may be a Kuiper belt.

Betelgeuse (Alpha Orionis)

A star in the constellation of Orion. It is a red giant with a **spectral classification** of M2. A variable star, it changes its brightness between apparent magnitudes $+0.1$ and $+1.2$.

Big Bang

The theory that proposes that the Universe came into being in an instantaneous event between 15 and 20 billion years ago. According to the theory, everything contained within the Universe was created in that initial event and, as time has passed, the Universe has expanded and its contents have evolved into the stars and galaxies of today. The theory uses the laws of physics to describe everything that occurred in the Universe following the **Planck time**, which ended when the Universe was 10^{-43} seconds old. After this, space and time became distinct from matter and energy. Matter and energy continued to be completely interchangeable until the Universe was 15 seconds old. At that time, the elementary particles that make up atoms became stable. Atoms could still not form, however, because collisions between photons and subatomic particles prevented them from binding together. The forces of nature evolved into **gravity**, **electromagnetism**, the **weak nuclear force** and the **strong nuclear force**. As the Universe expanded and temperature fell, atoms formed.

Big Crunch

The theoretical demise of a **closed universe**. If the Universe contains enough mass to halt the **Hubble flow** and reverse its expansion, all the matter contained within it will be drawn together by the force of its mutual gravitational attraction. This collapse will cause the matter to be concentrated in an ever-decreasing volume with densities and temperatures reaching those attained in the **Big Bang**. Some ambitious ideas suggest that, since a big crunch would match the conditions of the Big Bang, perhaps it would cause another phase of expansion and the Universe would begin again. This is known as the oscillating universe.

Big Dipper

An asterism within the Great Bear (Ursa Major), also known as the Plough. It is comprised of seven stars, two of which point to Polaris, the pole star. Five stars of the constellation are, with Sirius, members of a widely separated **star cluster**.

binary star

A stellar system composed of two stars that orbit one another about their common center of mass. The two stars are held together by the force of their mutual gravity. Binary stars are twins in the sense that they formed together out of the same interstellar cloud. Since they may have formed with different masses, however, they will evolve at different rates. As stars grow to become red giants, once nuclear fusion of helium begins in their cores, they can fill the space known as their **Roche lobes**. This means that matter can pass through the inner Lagrangian point, where the lobes touch, to form an **accretion disk** around the companion star. This, in turn, increases the mass of the companion and thus causes it to evolve faster.

Roughly one half of the stars in the sky are binaries or multiple systems. In some cases, the stars are so far apart that they can be clearly seen as having two components. These are known as visual binaries. Other binary stars are too close together for the separation between them to be detected directly. A star that displays a periodic "wobble" in its passage through space is being acted on by a force of gravity, caused by the mass of a companion star. So, although the second star cannot be seen, its presence can be inferred by the visible star's motion. These are called astrometric binaries.

If the motion caused by the companion star is too small to be detected by this method, **spectroscopy** can reveal its presence. The spectra of stars sometimes reveal the presence of two stars, either by containing incompatible spectral lines or by displaying movement of the lines caused by the **Doppler effect** as the stars orbit each other. These are known as spectroscopic binaries. Some binary stars appear to be variable stars. They are the eclipsing binaries, which have an orbital plane inclined to the Earth so that the stars pass in front and behind of one another, causing eclipses which dim the light output. Another, more violent variable, known as a **nova**, is thought to be caused within a binary star system in which one member is a white dwarf. *See also* **supernova**.

binding energy

The energy equivalent of the **mass defect** of an atomic nucleus.

bipolar outflow

A **stellar wind** that appears to leave a star along its polar axes in preference to all other latitudes. It often represents significant periods of mass loss in a star's life. They tend to occur during the **protostar** and **pre-main sequence** phase and, again, during the **red giant** phase just before the production of a planetary nebula. It is uncertain whether the bipolar flow is due to material not being ejected at other stellar latitudes or whether something, such as an **accretion disk**, blocks the material in the equatorial regions allowing only material ejected at the poles to escape. It has also been suggested that magnetic fields may constrain the outflowing material. The outflows carve out cavities in the surrounding interstellar medium and cause the formation of reflection nebulas.

black body radiation

The radiation emitted after a perfect absorber of radiation reaches a temperature higher than that of its surroundings. It covers the entire electromagnetic spectrum, with the peak emission taking place at a wavelength that depends on the emitting body's temperature. A perfect absorber does not reflect any electromagnetic radiation falling on it. All the radiation is absorbed and converted into internal energy, or heat, which is then re-radiated as black body radiation. Perfect absorbers are often referred to as black bodies because anything black tends to absorb radiation. The radiation emitted from stars approximates that of a black body. This is why the color of a star, which corresponds to its peak emission, can be used to determine its effective temperature.

black dwarf

The theoretical cold remnant of a dead, low-mass star. After a **red giant** has produced a planetary nebula and nuclear fusion has stopped in its core, the remains collapse to form a **white dwarf** star. The white dwarf is very hot and gradually radiates this heat energy into space. As it is no longer producing energy, once this residue has been radiated it becomes a cold, dark object known as a black dwarf. The cooling times for a white dwarf are so large, however, that the Universe may not yet be old enough to contain any black dwarfs. Detecting them is very difficult, because they are so small and radiate such small amounts of energy.

black hole

A stellar core remnant so dense that not even photons of electromagnetic radiation can escape from it. The most important property of a black hole is its mass. During the **Big Bang**, so-called primordial black holes were formed. These contained very little mass, perhaps only a few kilograms. At the other end of the scale, **active galactic nuclei** are thought to be powered by black holes with masses possibly thousands or millions of times greater than that of the Sun. A black hole is a region of space about which outside observers can never know anything, because information cannot escape its gravitational field. The spherical boundary at which information is lost forever is known as the event horizon. At the center of this region is the singularity: the point at which matter is compressed to an infinitely dense state.

In a rotating black hole, a region surrounding the event horizon is known as the ergosphere. Here the spacetime continuum is dragged around by the rotation of the black hole. The boundary of the ergosphere is called the stationary limit. As matter spirals down towards a black hole, it forms an **accretion disk**. The gases in these disks are so hot that they emit electromagnetic radiation in the form of X rays. It is these that allow astronomers to detect black holes. Black holes of stellar mass are thought to originate in **supernovae** explosions. If a **collapsar** composed of degenerate matter contains more mass than the Oppenheimer–Volkoff limit, it will become a black hole. *See* **Hawking radiation**.

blazar

The most active type of active galaxy. The name is a combination of **BL Lacertae** object and **quasar**; both can be described as blazars if they show violent variability in the optical region of the electromagnetic spectrum. According to some unified theories of **active galaxies**, the activity in blazars is caused by jets of gas being expelled from the active galactic nucleus, at speeds close to the speed of light, almost directly into the line of sight from Earth.

BL Lacertae

An **active galaxy**, the first blazar to be discovered. It was originally classified as a variable star in 1941 but is much more complex. In 1968 it was discovered to be a strong radio source. A class of active galaxy is now named after this object. BL Lacertae objects are thought to be the superbright nuclei of active elliptical galaxies. They contain no spectral lines in their spectra, which, theory suggests, is because there are no gas clouds surrounding the AGN (**active galactic nucleus**) in BL Lacertae objects; thus, no atomic absorption or atomic emission can take place. The electromagnetic radiation from the visible part of the spectrum is highly polarized and is thought to be **synchrotron** emission from electrons spiraling around magnetic field lines.

blue supergiant star

Any star in the upper left section of the **Hertzsprung–Russell diagram**. Blue stars have spectral classifications of O and B. Supergiant stars are many tens of times the mass of the Sun. They typically have absolute magnitudes of between $^-5$ and $^-10$.

boson

A particle that does not obey **Pauli's exclusion principle** – for example, a photon. Bosons can be identified because they have an integer **spin** quantum number.

Bok globule

A spherical dark nebula believed to be a region of interstellar clouds undergoing gravitational collapse. The globules appear as round clouds silhouetted by background stars or emission nebulas, and consist predominantly of gas. The temperature of the globules is about 10–15 kelvin, which is sufficiently low for the kinetic energy of the gas particles not to resist the collapse. As a globule collapses, it breaks up; the fragments collapse into **protostars**. Bok globules vary in mass, from a fraction of a solar mass to several hundred solar masses. They are named for a Dutch astronomer who observed the less massive variety early in the 20th century.

brown dwarf

A celestial object, formed by gravitational collapse, which is less than 0.08 solar masses. Below this, the temperature and pressure in the central regions of the object are insufficient to ignite the nuclear fusion of hydrogen into helium at its core. Brown dwarfs radiate energy into space but it is in the infrared region of the electromagnetic spectrum. The radiation carries away the residual energy left over from the formation of the brown dwarf. In appearance and composition, brown dwarfs may be very similar to the planet Jupiter. They are believed to exist in vast numbers throughout the Galaxy but because of their small size and meager energy output they are difficult to detect. Several possible brown dwarfs have been reported, though none has been confirmed.

bubble chamber

An instrument attached to an experimental particle accelerator that detects the passage of high-energy particles by the ionization they cause to the atoms in the chamber. As the high-energy particles ionize the atoms that they pass close to, energy is released which boils the bubble chamber liquid. This produces bubbles which show the path that the particle has taken. When the bubbles are large enough they are photographed and analyzed. The path characteristics indicate which type of particles created them.

Cancer

The Crab, a spring constellation in the Northern Hemisphere, the fourth sign of the **zodiac**. At the time the zodiacal system was adopted, Cancer marked the northernmost limit of the ecliptic. A hazy object near the center of Cancer is a cluster of stars named Paesepe, the Beehive.

Canis Major

A winter constellation known as the Great Dog. Canis Major is distinguished by Sirius, the brightest star in the sky. The constellation is easy to find because the three stars in Orion's belt point to it. Canis Major is close to Canis Minor.

Canopus

The brightest star in the Southern Hemisphere constellation Carina. Next to Sirius, it is the brightest star in the sky. Canopus is a red **supergiant** and has a **spectral classification** of F.

Capella

The brightest star in the constellation of Auriga. Detailed observations show that it is a spectroscopic binary. The main component is a G type giant and the companion is an F type dwarf. *See also* **binary star** and **spectral classification**.

Capricornus

The Sea Goat, a constellation of the Southern Hemisphere, lacking any bright stars; the 10th sign of the **zodiac**. Lying between the constellations of Aquarius and Sagittarius, Capricornus in ancient times lay at the southernmost limit of the **ecliptic**.

carbon cycle

In astronomy, a sequence of six **nuclear fusion** events in which a helium nucleus (two protons and two neutrons) is formed from four hydrogen nuclei (four protons) in a reaction using a carbon nucleus as a catalyst. The carbon is converted into **isotopes** of nitrogen and oxygen, before returning to its original form in the final stage of the reaction. Also called the carbon–nitrogen or carbon–nitrogen–oxygen (CNO) cycle, it takes place at temperatures greater than 4×10^6 Kelvin. It is thought that this is the main way that helium is produced in stars several times more massive than the Sun. *See also* **proton-proton chain**.

cD galaxy

A huge elliptical galaxy found at the center of a cluster of galaxies. The cD designation stands for "cluster dominating". These galaxies are believed to have grown so large because they have merged with other galaxies. Indeed, several cD galaxies have multiple galactic nuclei, as if they are still in the process of "cannibalizing" smaller galaxies.

CONNECTIONS

CLASSIFICATION OF GALAXIES **88**

INTERACTING GALAXIES **100**

---

celestial equator

The circle formed where a projection of the Earth's equator cuts the celestial sphere, along which right ascension and declination are measured.

celestial object

Any object, such as a star, galaxy or planet, that is projected onto the **celestial sphere**.

celestial sphere

The projection of space onto an imaginary sphere surrounding the Earth and centered on the observer. Only half of the celestial

---

CENTER OF MASS

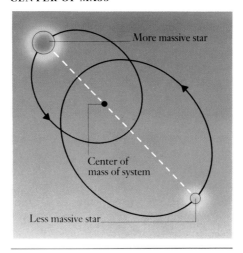

More massive star

Center of mass of system

Less massive star

---

sphere is ever visible to an observer because the other half is masked by the surface of the Earth. The celestial poles are points on the sphere directly above the Earth's poles, and the celestial equator is a projection of the Earth's equator. Several coordinate systems exist such as **altitude** and **azimuth**, **right ascension** and **declination**. The sphere is split up into arbitrary areas, the **constellations**.

center of mass

The point at which a system of masses would balance if placed on a pivot, and the point at which the mass of the object can for many purposes be thought to be concentrated. For example, in a binary star system, the center of mass is nearer to the more massive star than to the less massive component.

centrifugal force

The apparent force that pushes an object outward when it is in circular motion. The motion outward is experienced because of changing frames of reference. For instance, when a car is turning a corner, the car's

---

CEPHEID VARIABLE STAR

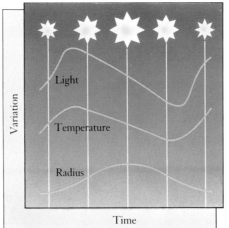

Light

Variation

Temperature

Radius

Time

---

frame of reference is changing. However, because the passengers do not expect to change their frame of reference, they continue traveling in a straight line and feel pulled or thrown to the outer side of the car.

Cepheid variable star

A star that has left the **main sequence** and begun to pulsate because it is unstable. During pulsation, the radius, temperature and luminosity vary within a regular, cyclical period of between 3 and 50 days, depending on the average **absolute magnitude** of the star. Temperature can vary by about 1500 kelvin while radius varies by about 10–30 percent. Luminosity changes by about one magnitude. Cepheids are named for the prototype star, Delta Cephei, and are generally yellow giant stars on their way to becoming **red giant** stars. They occupy a region on the **Hertzsprung–Russell diagram** known as the instability strip because of its elongated shape. There are two types of Cepheids variable star because of the two distinct stellar populations. Population I Cepheids are known as classical Cepheids and are, on average, two magnitudes brighter than W Virginis stars, which are Cepheids in **population II stars**.

CERN

A European-run laboratory complex, at which the physics of **subatomic particles** is studied. It contains a huge circular particle accelerator, known as the Super Proton Synchroton (SPS), one of the most powerful proton accelerators in the world. CERN also possesses the Large Electron Positron (LEP) collider, which performs the same function but for electrons and their antimatter counterparts, positrons.

Chandrasekhar limit

The mass limit at which the force of gravity overcomes the pressure produced by electron **degenerate matter**. At this mass limit, which corresponds to 1.4 times the mass of the Sun, the electrons are forced inside the atomic nucleus, where they combine with the protons to form neutrons. The gravitational collapse is then halted by the pressure exerted by the neutrons, because they are in a state of matter known as **baryon degenerate matter**. This pressure is sufficient to halt further collapse unless the object contains more than three times the mass of the Sun. *See also* **Oppenheimer–Volkoff limit**.

CONNECTIONS

COLLAPSING AND EXPLODING STARS **126**

NEUTRON STARS AND PULSARS **128**

---

chromosphere

The layer above the **photosphere** on a star. On the Sun, it extends upward about 2000–3000 km. At its base it has a temperature of 4500 kelvin, rising to 100,000 kelvin at the top, where it joins the corona.

closed universe

Any model of the Universe in which the ratio of the mean density of the Universe to the **critical density** is greater than one. Our universe is observed to be expanding. However, the expansion is slowed by the gravity of the matter within it. If there is enough matter for the expansion to be halted and then reversed, the Universe will collapse and a **Big Crunch** will be brought about. Observations suggest that the luminous matter in the Universe – for example stars and galaxies – provides only 10 percent of the material needed to make the mean density of the Universe equal to the critical density. In the revision to the standard model of the Big Bang known as **inflationary cosmology,** the density must equal the critical density. If this theory is correct, vast amounts of **dark matter** must make up the missing mass.

clusters of galaxies

Collections of galaxies that are held together by the force of their mutual gravity. Small clusters, with perhaps a few dozen members, are known as groups. Our own Galaxy, the Milky Way, is a member of a galaxy group known as the Local Group. The Andromeda galaxy (Messier 31) is also a member of this Local Group. Clusters contain hundreds or thousands of galaxies and, typically, have diameters of a few tens of millions of light-years. Different clusters have different shapes, for which two broad classifications exist. Spherical clusters contain predominantly elliptical galaxies and often have a **cD galaxy** at their center. It is thought that many **galactic mergers** have taken place within these clusters. Irregular clusters, in which the galaxies do not come into close contact with one another, contain predominantly spiral galaxies. In between the galaxies, the intergalactic medium is composed of very tenuous but very hot gas. This gas is detectable because it emits X rays.

collapsar

A star composed of degenerate matter, such as a white dwarf (made up of electron degenerate matter) or a neutron star (composed of baryon degenerate matter). **Black holes** are also classed as collapsars. The name refers to the fact that these objects have collapsed under gravity because energy is no longer generated within their cores.

color index

The difference in brightness of a star, as measured at two different wavelengths. The brightness is measured using **photometry** through different colored filters, which isolate specific wavelength ranges that can be very narrow or much wider. Careful choice of filters and measurements can yield a range of information about a celestial object.

comet

Any icy object that exists within the Solar System. Most comets are thought to lie in regions known as the Kuiper belt and Oort cloud. The Kuiper belt theoretically begins just outside the orbit of Pluto. It is shaped like a flared disk and gradually extends upward and downward to enclose the Solar System within a spherical shell of comets known as the Oort cloud. So far away from the Sun, and so small, comets are very difficult to observe. When they fall toward the Sun because of gravitation instabilities in their orbits, they can be more easily studied. As a comet approaches the inner Solar

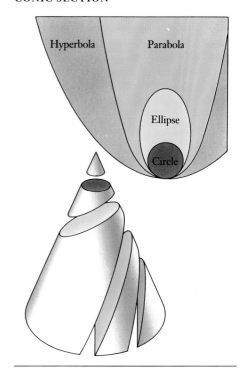

System, it receives increased amounts of solar radiation which sublime the ices on its surface, leading to the creation of a tail. There are two types of tail that a comet will develop: an ion tail (often referred to as a type I tail) and a dust (or type II) tail. The tails are formed because the subliming ices are throwing off jets of dust and gas into space. The gas is ionized (electrically charged) and is swept away by the **solar wind**. This means that the ion tail always points directly away from the Sun and thus appears straight. The dust tail is not shaped by the solar wind but is left in the comet's orbit. This causes the dust tail to be a sweeping fan-shaped object.

Some comets, such as Halley's comet, are short-period comets. These are thought to have originated in the Kuiper belt. They have highly elliptical orbits which bring them into the inner Solar System on timescales of up to 100 years or so. With each passage, they lose a little more of their mass until, eventually, they break up altogether. Streams of dust particles in comet orbits are responsible for meteor showers. Long-period comets are thought to fall inward from the Oort cloud. They often suffer gravitational perturbations from the giant planets, particularly Jupiter, which alter their orbits. In some cases, they can be converted to short-period comets; in others, their orbits are changed to a hyperbolic shape. These will then escape the Sun's gravitational influence and wander the depths of interstellar space.

COMET

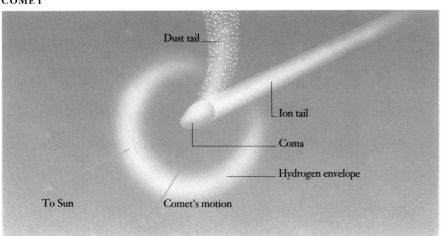

Dust tail

Ion tail

Coma

Hydrogen envelope

To Sun   Comet's motion

compact galaxy

Any galaxy that is considerably denser and smaller than other galaxies.

conic section

The generic term for circles, ellipses, parabolas and hyperbolas. The term derives from the fact that all of these shapes can be produced by taking a cone and "slicing" it in various ways.

constellation

An arbitrary grouping of stars that form a pattern; the term also refers to the area of the **celestial sphere** that bounds the star pattern. The total area of the celestial sphere is divided into 88 constellations. These vary in size and shape from Hydra, the sea monster, which is the largest at 1303 square degrees, to Crux, the cross, the smallest at 68 square degrees.

convective zone

The region within a star in which the predominant form of energy movement is by convection. Vast cells of hot gas become buoyant and rise through their surroundings, carrying energy with them. As they release this energy in the higher, less dense regions of the star, conditions change and the gas cell sinks back down into the deeper interior regions. In the Sun, the convective zone begins just below the **photosphere** and extends for a distance equivalent to one-fifth of a solar radius. In less massive stars than the Sun, the convective zone extends through much more than one-fifth of the star's radius. In high-mass stars, however, the convective zones are often much smaller and, in some cases, insignificant.

corona

The outer part of the Sun's atmosphere. *See also* **chromosphere, photosphere** and **Sun**.

cosmic background radiation

Weak electromagnetic radiation that is believed to be a remnant of the radiation released during the decoupling of matter and energy, several hundred thousand years after the **Big Bang**. The peak wavelength of the emission is in the **microwave** region of the electromagnetic spectrum. The emission profile corresponds to a **black body radiation** curve of temperature 3 kelvin. When this radiation was released, it was very much hotter but, as it has traveled through space to reach the Earth, it has been redshifted enormously. The discovery of this radiation, which was announced by Arno Penzias and Robert Wilson in 1965, provided great support to the Big Bang model of cosmology. The background radiation appears to be isotropic: it comes from all parts of the sky with equal intensity. Slight variations in the intensity of the background radiation have been detected using the satellite COBE (Cosmic Microwave Background Explorer). The variations are very small, typically only one part in a hundred thousand, but they show that matter in the early Universe was not uniformly distributed. It is these ripples that formed the seeds of the galaxy clusters we see throughout the Universe today.

CONNECTIONS

NATURAL HISTORY OF THE BIG BANG **78**

INFLATIONARY UNIVERSE **80**

INFANT UNIVERSE **82**

cosmic ray

A subatomic particle, usually a proton, moving through space at close to the speed of light. How cosmic rays are formed and accelerated is uncertain. They may be formed entirely within our own Galaxy, although many astronomers believe that some have an extragalactic origin. The direction from which these rays strike the Earth cannot be used to analyze their origin because cosmic rays are charged particles and so they are deflected by the magnetic fields pervading the Galaxy. When a cosmic ray strikes an atomic nucleus in the Earth's atmosphere, it produces a secondary cosmic ray. This is in the form of a cascade of subatomic particles, such as mesons. It is the secondary cosmic rays that are observable from the surface of the Earth.

cosmological principle

A cornerstone in every cosmological model of the Universe: it states that there are no preferred vantage points in the Universe. In other words, from the Earth we see everything that could be seen from any other place within the Universe. This principle implies that astronomical theories should apply to the Universe as a whole.

cosmology

The study of why the Universe is like it is, how it came into being, how it will evolve and how it will end. Cosmologists study the Big Bang, and consider the possibility of the existence of alternative universes with different laws of physics.

critical density

The average density of matter required to halt the expansion of the Universe. According to present estimates, the luminous matter in the Universe provides up to 10 percent of the required material. **Dark matter** is believed to make up the rest. *See also* **closed universe**.

CONNECTIONS

FATE OF THE UNIVERSE **132**

OPEN, FLAT OR CLOSED? **134**

LONG-TERM FUTURE **136**

Cygnus X-1

A source of intense X rays which is thought to be the site of a stellar-sized **black hole**. The X-ray source is a **binary star system** in which one star is a blue supergiant and the other is very dense and compact. The X rays are thought to be coming from hot gas in an **accretion disk** around the companion. The compactness of the companion suggests it is even more dense than a **neutron star**. Astronomers suggest it may be a black hole.

dark matter

A form of matter that is thought to exist in the Universe in vast quantities. It is difficult to detect because it is either nonluminous or

CONSTELLATION

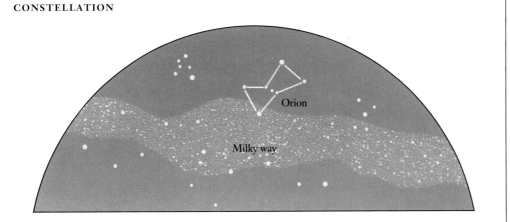

Orion

Milky way

of very low luminosity. Cosmologists have suggested that it should exist in order to explain various gravitational effects, specifically within galaxies and clusters of galaxies. In a spiral galaxy, the stars move as if large quantities of invisible matter exist around the disk of the galaxy. In clusters of galaxies, the individual galaxies move as if ten times as much matter exists as can be seen in the stars and the emission nebulas.

Dark matter is postulated to be of two main types: baryonic and exotic. Baryonic matter makes up the luminous portions of the Universe. Dark baryonic matter is made up of atoms of the familiar chemical elements which are simply bound into intrinsically low-luminosity objects such as planets, **brown dwarfs** and **black holes**.

Exotic dark matter may be cold dark matter or hot dark matter. Hot dark matter is composed of particles such as **neutrinos**. They are called hot because they travel nearly at the speed of light. It is uncertain whether neutrinos have any mass. If not, they cannot provide the gravity necessary to act as the dark matter. Cold dark matter comprises weakly interacting massive particles (WIMPs). These have relatively large masses, travel relatively slowly and interact

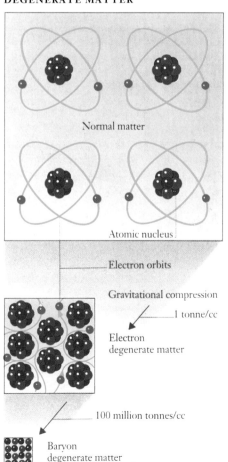

DEGENERATE MATTER

Normal matter

Atomic nucleus

Electron orbits

Gravitational compression

1 tonne/cc

Electron
degenerate matter

100 million tonnes/cc

Baryon
degenerate matter

only weakly with "normal" baryonic material. Many experiments have attempted to detect neutrino mass and WIMPs, but no conclusive evidence has yet been found.

dark nebula

A region of space in which the density of material is greater than that of the **interstellar medium**. Dark nebulas are visible because they block out the light from background stars and emission nebulas. It is within the dense cores of dark clouds that stars are formed. *See also* **absorption nebula** and **Bok globule**.

de Broglie wavelength

The calculated wavelength of a particle that is being considered as a wave, according to the principles of **wave-particle duality**. The amplitude of the wave at a given location represents the probability that the particle is also there, in accordance with **Heisenberg's uncertainty principle**.

declination

An angular measurement, similar to altitude, used to reference celestial objects. It is measured from the celestial equator, the projection of the Earth's equator on to the **celestial sphere**.

decoupling of matter and energy

An era in the early Universe, between 300,000 and 500,000 years after the **Big Bang**, when the temperature fell sufficiently for electrons to be captured by atomic nuclei. Before this, the electrons existed as a random distribution of particles throughout space. Photons of **electromagnetic radiation** frequently collided and scattered off in new directions. After the decoupling, the electrons were held close to the atomic nuclei by the force of electromagnetism, leaving large volumes of empty space through which the photons could travel. Before decoupling, space was opaque to radiation, but afterward it became transparent to it. This event is observable today as the **cosmic background radiation** which was released during the decoupling. Decoupling ended radiation's dominance over matter and allowed gravity to begin shaping the Universe.

CONNECTIONS

NATURAL HISTORY OF THE BIG BANG **78**
BEGINNINGS OF STRUCTURE **84**

degenerate matter

Matter in which the usual atomic structure has broken down because of the weight of overlying material. There are two forms of

degeneracy. Electron degenerate matter occurs when the electrons are forced to try to occupy the same quantum states. This does not happen, as described by **Pauli's exclusion principle**, and so the electrons exert a pressure that resists further collapse. At higher mass limits, the electrons are forced into the atomic nuclei where they combine with the protons to form neutrons. This produces baryon degenerate matter, in which the neutrons begin to exert resistive pressure. *See also* **Chandrasekhar limit**.

CONNECTIONS

COLLAPSING AND EXPLODING STARS **126**
NEUTRON STARS AND PULSARS **128**

dense core

A region within a **dark nebula** that has almost reached the density that will cause it to undergo gravitational collapse and form stars. *See also* **Bok globule**.

density

The ratio of an object's mass to its volume.

density wave theory

A theory to explain the formation of stars in the arms of spiral galaxies, based on the hypothesis that the interstellar medium is compressed into regions of higher density along the leading edges of the spiral arms. These regions cause dense cores to be produced in **dark nebulas**, which collapse to form stars. As the stars are born, **OB associations** (regions of high-mass stars) are created, which advance the density wave by pushing the interstellar medium onward by stellar winds. The density wave advances around the galactic nucleus, triggering a new wave of star formation. Thus the spiral arm appears to move as the new stars begin to shine.

DENSITY WAVE THEORY

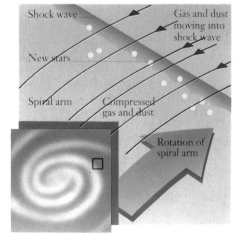

Shock wave

Gas and dust moving into shock wave

New stars

Spiral arm

Compressed gas and dust

Rotation of spiral arm

deuterium

An isotope of hydrogen which contains one neutron and one proton in its atomic nucleus.

Doppler effect

A physical process that alters the wavelength of electromagnetic radiation or sound because the source of emission and the observer are in relative motion. When the source and observer move closer together, the wavelength is "squashed". For sound waves this means that the pitch increases. For electromagnetic waves it means that the photons become more energetic. When the source and the observer are moving apart, the wavelength is "stretched". Sound drops in pitch and photons are shifted to the less energetic, red end of the electromagnetic spectrum. This gives rise to the term **red-shift**. Redshift is easily observable by looking at the light from distant galaxies. As the Universe is expanding, nearly every galaxy is moving away from our own. As a result, the electromagnetic radiation released by galaxies is redshifted before it reaches the Earth.

double stars

Any pair of stars that appear close to one another on the celestial sphere. Double stars may be physically linked to one another by gravity, in which case are known as **binary stars**, or simply chance alignments.

dwarf elliptical galaxy

Any elliptical galaxy that is substantially smaller than the average.

dwarf irregular galaxy

Any irregular galaxy that is substantially smaller than the average.

dwarf star

A star lying on the **main sequence** that is too small to be classified as a giant star or a supergiant star. For instance, the Sun is a yellow dwarf star. Main sequence stars of **spectral classification** K and M are known as red dwarf stars. Celestial objects that have formed in the same way as stars but have not reached the mass limit required to ignite nuclear fusion, are known as brown dwarfs. The cores of dead stars, which are composed of electron **degenerate matter**, are known as white dwarfs.

DOPPLER EFFECT

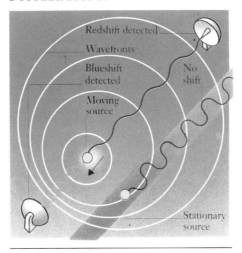

Redshift detected
Wavefronts
Blueshift detected
No shift
Moving source
Stationary source

Earth

The third planet in distance from the Sun. The Earth is the home of humankind. It is a member of the four terrestrial planets of which the inner Solar System is composed. All these worlds are rocky bodies that have high densities and tenuous (or even non-existent) atmospheres. The Earth is unique in that it exists under conditions that allow water to take the forms of a solid, a liquid and a gas. In fact, 70 percent of the surface of the planet is covered by liquid water oceans. At the center of the planet is a solid core of nickel and iron. This has a radius of about 1300 km. Surrounding this the conditions are such that the nickel and iron exist as a liquid. This region, the outer core, has a thickness of about 2200 km. Encompassing these cores, for a distance of just under 3000 km, is the region known as the mantle. This is composed of rocky material, which gives way to the crustal rocks about 30 km from the surface of the Earth. The volume of the Earth's atmosphere is composed of 78 percent nitrogen, 21 percent oxygen and 1 percent of the other atmospheric constituents. The Earth possesses a magnetic field, which is believed to be generated in the outer core by the motion of the liquid metal. It also possesses one natural satellite, the Moon.

eclipse

A chance alignment between the Sun and two other celestial objects within the Solar System in which one body blocks the light of the Sun from the other. In effect, the outer object moves through the shadow of the inner object. Lunar eclipses occur when the Earth passes directly between the Moon and the Sun. Solar eclipses occur when the Moon passes directly between the Sun and the Earth. If its alignment is not perfect, there is a partial solar eclipse. When the Moon passes directly between the Earth and

the Sun but it is at the farthest point of its orbit, so it does not quite cover the Sun, an annular eclipse occurs. In this case, an annulus of sunlight remains visible around the silhouette of the Moon. Eclipses can also occur outside the Solar System in **eclipsing binary** star systems.

eclipsing binary

A binary star system in which chance alignment of the orbits of the two stars causes them to pass, periodically, in front of one another and produce eclipses. As a result, the combined light of the stars varies, giving the binary the appearance of a variable star. An example is Algol in the constellation of Perseus. The two stars in this system orbit one another with a period of 2.87 days. Algol is visible to the naked eye, and when the system is in eclipse it is noticeably dimmer in the night sky.

ecliptic

The ecliptic is the path that the Sun takes across the sky every day. It is the plane of the Solar System, projected onto the **celestial sphere**. The planets, as viewed from the Earth, always stay close to the ecliptic in their passages across the sky. The zodiacal constellations all lie on the ecliptic.

effective temperature

Of a star, the temperature required for a black body to have the same luminosity per unit area as the star. *See also* **black body radiation**.

electromagnetic force

See **electromagnetism**.

electromagnetic radiation

A form of **energy** that propagates through space as oscillating electric and magnetic disturbances. Electromagnetic radiation travels through a vacuum at the speed of light. Gamma rays, ultraviolet radiation, visible light, X rays, infrared radiation, microwaves and radio waves are all electromagnetic radiation. Usually thought of as a wave phenomenon with wavelength and frequency, electromagnetic radiation can also be considered as being carried by a particle known as a **photon**. Electromagnetic radiation is produced when interactions occur through the force of electromagnetism.

electromagnetic spectrum

The entire range of electromagnetic radiation, from the longest wavelength radiation – radio waves – through microwaves, infrared rays, visible light rays, ultraviolet rays, X rays to the shortest wavelength radiation – gamma rays. Although all the forms of electromagnetic radiation propagate through space at the speed of light, the shorter-wavelength radiations carry more energy per **photon** than the longer.

electromagnetism

One of the four **fundamental forces** of nature. Electromagnetism acts between all particles that carry an electric charge, that is, electrons, protons and atomic ions. It is manifested in two interconnected forces – electricity and magnetism – both of which have polarities as well as magnitudes. In electric charges, the polarities are distinguished by sign, positive or negative. In magnetism, the polarities are denoted as north or south. In both cases, like polarities repel and unlike polarities attract. The **virtual particles** that carry the electromagnetic force are virtual photons.

CONNECTIONS

FORCES AND FIELDS **68**

UNIFYING THE FORCES **70**

NATURAL HISTORY OF THE BIG BANG **78**

electron

One of the elementary particles of nature, also called a subatomic particle. Electrons belong to the lepton family and are the negatively charged components of atoms. In the simplest model of the atom, electrons occupy specified orbits around the atomic nucleus. When electrons absorb energy from passing **photons**, they jump between the energy levels surrounding the atomic nucleus. This is known as atomic absorption. When the electrons jump back down to their original energy levels, atomic emission takes place, releasing a photon of electromagnetic radiation. If an electron absorbs enough energy it escapes from the atom altogether. This is called ionization. Electrons were the first elementary particle to be discovered (as a constituent of cathode rays); its antiparticle, the positron, was the first antiparticle to be discovered.

CONNECTIONS

INSIDE THE ATOM **50**

THE QUANTUM VIEW **64**

EXCLUSION AND UNCERTAINTY **66**

electron degenerate matter

A form of degenerate matter in which the weight of the overlying material tries to force all of the electrons surrounding the atomic nucleus into the lowest energy quantum state. They cannot do this because of **Pauli's exclusion principle** and so the electrons exert a pressure (electron degenerate pressure) which halts further collapse. This pressure is sufficient up until 1.4 solar masses, the **Chandrasekar limit**. At this point gravity overwhelms the degeneracy pressure and a further collapse takes place. *See also* **baryon degenerate matter**.

electron shells

Groupings of electrons around atomic nuclei. They are defined by the laws of **quantum theory** and can contain two or more quantum states. Electrons then take up positions in these quantum states. The concept is a more advanced version of the ideas about electron orbits.

electroweak force

The combined force of **electromagnetism** and **weak nuclear force**. The unification of these two forces is described by the Glashow–Weinberg–Salam theory. It states that, at sufficiently high energy, the forces of electromagnetism and the weak nuclear force act in exactly the same way. The theory has been verified as true in large particle-accelerators and is the first step in unifying all the forces of nature. The next step combines the **strong nuclear force**. This is known as a **grand unified theory**. The further unification of gravity would first require a quantum theory of gravity.

CONNECTIONS

UNIFYING THE FORCES **70**

NATURAL HISTORY OF THE BIG BANG **78**

INFLATIONARY UNIVERSE **80**

element

A specific atomic type. All atoms of the same element have the same number of protons in their atomic nucleus. Elements are grouped by atomic number and physical and chemical characteristics on the **Periodic Table**.

elementary particle

Any one of five stable particles that make up atoms, molecules and all matter in the Universe. The elementary particles are electrons, protons, neutrons, neutrinos and photons. A sixth particle, the graviton, is thought to exist but has not yet been officially discovered and identified in experiments. Protons, neutrons and electrons make up all visible material in the Universe; neutrinos, photons and gravitons carry energy generated by interactions between the others. Elementary particles are subatomic particles but are not the same as **fundamental particles**. *See also* **atom, fundamental forces** and **quark**.

CONNECTIONS

INSIDE THE ATOM **50**

PARTICLE EXPERIMENTS **52**

THE PARTICLE FAMILY **62**

elliptical galaxy

A galaxy with an ellipsoidal shape. Elliptical galaxies are characterized by an absence of spiral arms. Most are deficient in dust and few possess regions of current star formation. From observations, 80 percent of galaxies are elliptical. One theory suggests that elliptical galaxies are produced by galactic mergers of spiral galaxies. *See also* **cD galaxy**.

emission nebula

A cloud of interstellar gas that glows because it absorbs, and then re-emits, electromagnetic radiation from hot stars embedded within it. The ultraviolet radiation from these embedded stars (usually of **spectral classification** O and B) ionizes or excites the hydrogen in the cloud (these clouds are sometimes called HII regions). The predominant wavelength of the radiation released by the hydrogen gas is 656 nm, which is in the red end of the visible spectrum; thus emission nebulas are a characteristic pinkish-red color. As the solar winds from the ionizing star blow away the central regions, the nebula dissipates. Emission nebulas are usually isolated regions of much larger interstellar clouds.

CONNECTIONS

ON THE MAIN SEQUENCE **122**

COLLAPSING AND EXPLODING STARS **126**

energy

The amount of "work" an object or a system can perform. Some common forms of energy are kinetic energy, which is exhibited by a moving object; potential energy, which is possessed by an object suspended in a gravitational field; and electromagnetic energy, which is carried by photons. Einstein's theory of special relativity shows that energy is interchangeable with mass ($E = mc^2$).

energy level

The energy associated with a **quantum state**. Electrons can exist only around atoms with certain fixed quantities of energy. These quantities are known as energy levels and are defined in accordance with **quantum theory**.

equatorial mounting

A mounting for a telescope, designed so that the two axes supporting it are aligned, one to the polar axis and the other to the equator of the Earth. This allows use of the **right ascension** and **declination** coordinate system. There are many variations on the designs of equatorial mounting but they all rely on the principle that to counteract the rotation of the Earth, the telescope needs to be driven only around the polar axis. This method of mounting is expensive, so the simpler and less expensive **altazimuth mounting** continues to be used by some astronomers, though it is not as sophisticated.

equinox

One of two points where the **ecliptic** intersects the **celestial equator**.

ergosphere

The region around a **black hole** in which the **spacetime continuum** is dragged around by the rotation of the black hole.

EQUATORIAL MOUNTING

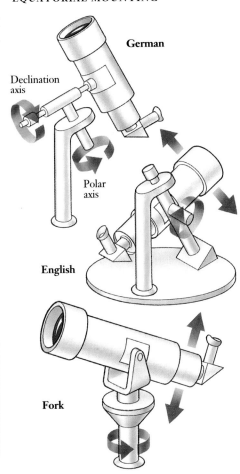

German

Declination axis

Polar axis

English

Fork

Here, not even frames of reference that are at rest with respect to the spacetime continuum are stationary with respect to the rest of the Universe. *See also* **stationary limit**.

eruptive variable

Any star that increases its brightness unpredictably. Examples include novae and super-**novae** as well as **red dwarf** stars, which can suffer violent stellar flares. *See also* **flare star**.

ESA (European Space Agency)

A collaborative body, consisting of 12 European countries, which designs and constructs space missions such as the Ariane satellite launchers. In the early 1990s ESA was developing space science missions to study the **solar wind**.

escape velocity

The velocity that an object must attain if it is to escape a planet's gravitational influence. The escape velocity for an object to leave the Earth is approximately 11 km/sec.

event

A point on a four-dimensional continuum which is referenced by three coordinates of space and one coordinate to represent time.

event horizon

The spherical boundary of a **black hole** beyond which no information can escape because of the gravitational pull. *See also* **ergosphere** and **Schwarzschild radius**.

extraterrestrial

Any object or form of radiation that does not originate on Earth.

fermion

A proton, neutron, electron or any other **subatomic particle** that obeys **Pauli's exclusion principle**. Fermions all have half integer **spin**. They can be divided into two groups: hadrons are built from quarks and leptons are not. *See also* **boson** and **quark**.

flare star

A **red dwarf** star that undergoes rapid and unpredictable increases in brightness, which are thought to occur just above the star's photosphere. Flares are energetic events, caused by magnetic fields releasing energy. Flares are also observed regularly on the Sun

but do not have such a dramatic effect on its appearance. This is because the Sun is intrinsically much brighter than a red dwarf star, where any increase in brightness is therefore more noticeable. The energy released in flares is so great that it brightens the whole red dwarf star. Flares often die down after only a few minutes.

flatness problem

The puzzle arising from observations suggesting that the Universe is neither open nor closed but somewhere between the two ("flat"), in spite of an infinite number of possibilities that it could be open, and an equal number of possibilities that it could be closed. Possible answers to this problem are provided by the theory of **inflationary cosmology**. *See also* **"flat" universe**, **closed universe** and **open universe**.

"flat" universe

Any model of the Universe in which the ratio of the mean density of the Universe to the critical density is equal to one. The expansion of the universe is gradually being slowed by the force of gravity created by the matter within it. In the case of a flat universe, the expansion will be halted after an infinite length of time. *See also* **closed universe** and **open universe**.

force

A physical quantity that, when applied to a object, will induce it to change its state of motion or its frame of reference. Forces can also be applied to objects so that they deform them without inducing an overall motion.

force of nature

Any of the four natural physical forces that shape the Universe – strong nuclear force, weak nuclear force, electromagnetic force or gravity. *See also* **fundamental forces**.

frame of reference

A position within the spacetime continuum from which measurements and observations can be made. According to Einstein's theory of **special relativity**, frames of reference can also be moving through the spacetime continuum with a constant velocity. If an object is either accelerating or decelerating it is changing its frame of reference. Such accelerated frames of reference are dealt with by Einstein's theory of **general relativity**.

frequency

A measure of the number of wavelengths that pass a specific point in space within a specific time.

fundamental forces

Any of the forces by which elementary particles interact with one another. The four fundamental forces that shape the Universe are **gravity**, **electromagnetism**, the **strong nuclear force** and the **weak nuclear force**. Although they are distinct from one another in the present-day Universe, they are all thought to be slightly different aspects of the same **superforce**. This superforce is predicted by unified field theories and is supposed to have been active only in the first few instants following the **Big Bang**. As the Universe expanded and cooled, so the superforce broke up into the four fundamental forces known today.

Steps toward proving this theory begin with the unification of electromagnetism and the weak nuclear force to provide the observed **electroweak force**. The unification of the strong nuclear force and the electroweak force then provides the hypothetical **grand unified force**. If a quantum theory of gravity were to be developed, gravity too could be unified to produce the unified field superforce.

CONNECTIONS

FORCES AND FIELDS 68

THE BIG BANG 74

NATURAL HISTORY OF THE BIG BANG 78

INFANT UNIVERSE 82

fundamental particle

A subatomic particle that cannot be broken down into smaller components. Fundamental particles include quarks, gauge bosons and leptons (electrons, muons, neutrinos and tau particles), none of which have an internal structure consisting of smaller particles. Fundamental particles are not the same as elementary particles (electrons, neutrons, protons, neutrinos, photons and gravitons), which are made up of fundamental particles. *See also* **atom**, **elementary particle**, **fundamental forces**, **quark** and **subatomic particle**.

fusion

The process that occurs when atomic nuclei are squeezed so close together that their constituent protons and neutrons combine to form a single, heavier nucleus. The simplest form of fusion occurs when hydrogen nuclei fuse to form helium. Energy is given out in this fusion process, in accordance with

Einstein's equation $E = mc^2$ where m is the **mass defect**. It is the nuclear fusion of hydrogen atoms into helium that provides the energy in stars. *See also* **proton-proton chain**.

CONNECTIONS

INSIDE THE ATOM 50

KINDS OF STARS 102

STARS AND GALAXIES 104

VARIABLE STARS 114

ON THE MAIN SEQUENCE 122

galactic center

The central region of a galaxy. In our Galaxy, the Milky Way, the center lies in the direction of the constellation Sagittarius and is dominated by a radio source apparently at the precise center of the galaxy. This is known as Sagittarius A. The galactic center cannot be seen using optical telescopes because there are too many interstellar clouds in the **galactic disk** for light to be able to penetrate. The galactic center is observed, instead, using radio waves and infrared telescopes. A black hole may exist at the center of our Galaxy but as yet this is unproved. Some other galaxies possess an **active galactic nucleus (AGN)**.

galactic corona

A spherical region, with a radius of about 250,000 light years, that exists around our own and other spiral galaxies. The corona is thought to consist of **dark matter**.

galactic disk

The plane in which the spiral arms of **spiral galaxies** or **barred spiral galaxies** exist. Lenticular galaxies also possess disks but no spiral arms.

galactic halo

The spherical region around our own and other spiral galaxies consisting of dim stars, **brown dwarfs** and the stellar collections known as globular clusters. Its radius is that of the parent galaxy but as it is a sphere instead of a disk, its volume is much greater.

galactic merger

The process that takes place when two galaxies "collide" with one another. If they catch in one another's gravitational fields, they do not simply make a close pass and travel off again into intergalactic space, but spiral into one another, forming a single galaxy. **Elliptical galaxies** are thought to be the products of mergers between **spiral galaxies**. As the spaces between the individual stars in galaxies are so large, although the

galaxies appear to collide, the stars within them do not. The huge gravitational fields of the galaxies, however, distort their shapes and the close stellar encounters swing the stars into randomly orientated orbits. The interstellar clouds may collapse in the merger and many new stars may be formed. This process is called a starburst. As the galaxies become closer, long tails of stars are strung out behind them. These fragments may contain enough matter to be considered dwarf irregular galaxies themselves.

galactic nucleus

The central bulge of older stars found in the center of all spiral galaxies, including our own. The stars in this region are of stellar Population II. It is from the nucleus that the spiral arms emanate. The nucleus surrounds the **galactic center**.

CONNECTIONS

GALAXIES AND QUASARS 86

STARS AND GALAXIES 104

galaxies, classification of
See **Hubble tuning-fork diagram**

galaxy

A collection of stars, dust and gas that is held together by the force of its constituents' gravity. Galaxies occur in a wide variety of shapes and sizes. One of the most common

GAMMA-RAY ASTRONOMY

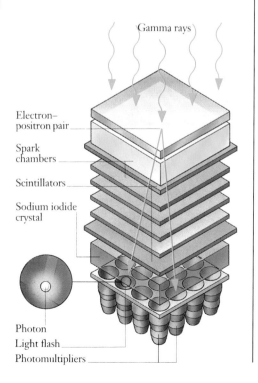

Gamma rays

Electron–positron pair

Spark chambers

Scintillators

Sodium iodide crystal

Photon

Light flash

Photomultipliers

classification systems for them is shown on the **Hubble tuning fork diagram**. This divides them into spiral, barred spiral, lenticular and irregular shapes. The smallest galaxies may contain fewer stars than a typical globular cluster, perhaps a few hundred thousand, whereas the largest include thousands of billions of stars. The smallest galaxies are the dwarf irregular and dwarf elliptical galaxies. The largest are the vast elliptical galaxies, especially the **cD galaxies** that are found at the center of a cluster of galaxies. Elliptical galaxies contain mostly old stars and little gas. There are also peculiar galaxies in which some form of energetic process is taking place in the galactic center. These are known as the **active galaxies,** and they come in a variety of types. Galaxies are held together in associations known as **clusters of galaxies**.

CONNECTIONS

Galaxy, the
See **Milky Way**.

gamma-ray astronomy
The study of gamma rays emitted by celestial objects. Because the wavelengths of gamma rays are so short, researchers often use special telescopes, known as **grazing incidence telescopes**, or other electronic detectors, to observe them. Of particular interest are gamma-ray bursts of up to 10 seconds which occur several times a year from sources widely distributed over the sky. Some 20 gamma-ray stars have been identified in the Milky Way.

gamma rays
The highest energy form of electromagnetic radiation. The wavelength of gamma rays is typically less than 10^{-12} m. *See also* **gamma-ray astronomy**.

gas giant
A type of planet made up of vast quantities of gas. In the Solar System, there are four gas giant planets: Jupiter, Saturn, Uranus and Neptune. Any rocky cores these planets do possess are very small compared with the amount of overlying gaseous material.

gas (in space)
See **interstellar medium**.

Gemini
The Twins, a constellation on the ecliptic named for its two brightest stars, Castor and Pollux. The third sign of the **zodiac**, Gemini gives its name to the Geminid meteor shower.

general relativity
An extension of Albert Einstein's previous work on **special relativity**, general relativity sought to deal with accelerating frames of reference. This led to the principle of equivalence, which states that, over small regions of the **spacetime continuum**, observers cannot tell whether they are being acted on by gravity or a uniform acceleration. By dealing with accelerating frames of reference, the theory of general relativity provides astronomers with the best theory to predict the effects of gravity. The general theory of relativity was published in 1915, and led on to the prediction that space could be shown to be curved and that light bends as it passes massive objects.

CONNECTIONS

geodesic
The shortest path between two points on the four-dimensional framework of the **spacetime continuum**. An object traveling solely under the influence of gravity follows geodesics.

giant molecular cloud
A vast area of molecular hydrogen that stretches through space for several hundred light-years. Such clouds are the largest single objects in our Galaxy and can contain up to 10 million solar masses of material. Isolated

GIANT MOLECULAR CLOUD

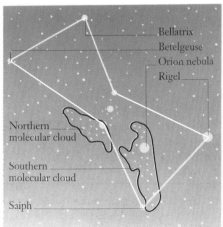

Bellatrix
Betelgeuse
Orion nebula
Rigel
Northern molecular cloud
Southern molecular cloud
Saiph

GLOBULAR CLUSTER

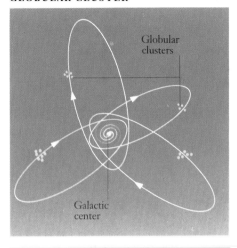

Globular clusters
Galactic center

regions of the giant molecular clouds have already undergone gravitational collapse, formed stars and are shining as emission nebulas.

CONNECTIONS

giant star
Any star that possesses a radius greater than ten times that of the Sun. Blue giant stars are members of the main sequence on the **Hertzsprung–Russell diagram**, whereas **red giants** are more evolved and burn helium in their cores.

globular cluster
A spherical collection of old stars found within galactic haloes of spiral galaxies and, more commonly, around elliptical galaxies. The globular clusters formed at the same time as the host galaxy. All the stars are deficient in metals and belong to stellar Population II, the same as those in the galactic nucleus of spiral galaxies. As they orbit the galactic nucleus, they periodically pass through the **galactic disk**.

grand unified force
The unified force that encompasses the strong nuclear force, the weak nuclear force and electromagnetism. *See* **grand unified theory**.

grand unified theory (GUT)
A theory that attempts to show that the strong nuclear force, the weak nuclear force and electromagnetism are slightly different aspects of the same **fundamental force**: the

grand unified force. The first step to GUT was the unification of the weak nuclear force and electromagnetism into the electroweak force. The **electroweak force** can be proved in particle accelerators, but to prove GUTs by this method would require collisions of higher energy than can currently be envisaged. Proponents of GUTs need to demonstrate the existence of a virtual particle that can transform quarks, the fundamental particles that make up protons, into leptons. The existence of such a particle would imply that protons are not stable, as had previously been thought, but decay (though at an extremely slow rate). Several experiments to observe proton decay have been in progress since the early 1980s, with little success so far. To provide a **unified field theory**, gravity must then be incorporated with the grand unified force.

CONNECTIONS

NATURAL HISTORY OF THE BIG BANG **78**

INFLATIONARY UNIVERSE **80**

FATE OF THE UNIVERSE **132**

---

gravitational lens

Any massive celestial object, such as a galaxy, that distorts the **spacetime continuum** so that the light from more distant objects is bent. This bending results in more than one image of the background object being visible to the foreground observer. The amount of bending allows an estimate to be made of how massive the gravitational lens is. Such calculations suggest the existence of **dark matter.**

graviton

The hypothetical **elementary particle** of **gravity**. There is no technology to detect such particles, so its existence is unproven.

graviton decoupling

A theoretical epoch, early in the history of the Universe, in which a density was reached such that gravitons no longer constantly interacted with other particles. The gravitons became free to travel large distances, unhindered. *See also* **decoupling of matter and energy** and **neutrino decoupling**.

gravity

One of the forces of nature. It is an attractive force exerted between two or more objects that have mass. Despite being the weakest force of nature, gravity shapes the Universe on the largest of scales. Unlike electromagnetism, gravity only attracts and does not repel. **General relativity** is the theory that best explains the effects of gravity.

grazing incidence telescope

A telescope used to study high-energy electromagnetic radiation, for example X rays and gamma rays. Whereas most telescopes use a focusing device formed from the bowl of a three-dimensional parabola (a paraboloid), the mirror of a grazing incidence telescope consists of a ring from higher up on the wall of the paraboloid. This means that photons strike it at acute angles and are deflected, onto the detector.

group of galaxies

A small cluster of galaxies, numbering some tens of galaxies. The Milky Way is a member of the **Local Group**.

hadron

A subatomic particle composed of quarks. It can be either a meson (quark doublet) or baryon, such as a proton (quark triplet).

half-life

The time taken for half the nuclides (atoms) in a radioactive substance to decay.

Hawking radiation

The process by which a black hole can apparently evaporate into space. When virtual particles are produced near a black hole, it is possible for one member of the matter-antimatter pair to be pulled into the black hole while the other escapes into space. The particle that falls into the black hole negates some of its mass and so the black hole shrinks a little. To an observer it would appear as if the particle that escaped into space has come from the black hole and that the black hole is evaporating. The theory was devised by the Cambridge University professor and astronomer Stephen Hawking and is the subject of observational searches. However, no direct evidence for this process has yet been found.

HERBIG–HARO OBJECT

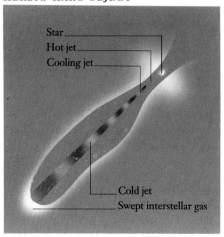

Star
Hot jet
Cooling jet
Cold jet
Swept interstellar gas

Hayashi track

The path on a **Hertzsprung–Russell diagram** that is followed by a protostar on its way to becoming a **main sequence** star.

heat

The internal energy of an object.

heat death

The ultimate fate of an open, expanding universe. It will occur if the contents of the universe reach thermal equilibrium; that is, everything is at the same temperature. If this happens, chemical reactions will cease. Stars, planets and galaxies will no longer exist and the universe will consist of a uniform "sea" of elementary particles. If grand unified theories are correct and the process of proton decay takes place, neutrons and protons will decay and the universe will become a diffuse sea of leptons. It may be that a closed universe could reach this condition before collapsing into a **big crunch**.

Heisenberg's uncertainty principle

A consequence of wave-particle duality, named for the German physicist Werner Heisenberg, this principle states that it is impossible to know simultaneously both the precise location of a particle and the precise direction in which it is traveling. The more precisely the position is known, the less precisely are the details of its motion. The uncertainty principle also permits the existence of virtual particles.

CONNECTIONS

EXCLUSION AND UNCERTAINTY **66**

FORCES AND FIELDS **68**

---

helium

The second element in the **Periodic Table**. It is characterized by possessing two protons in its atomic nucleus. Helium accounts for just under one quarter of the atomic matter in the Universe and was principally formed during the first few minutes after the **Big Bang**. It is also produced by nuclear fusion at the center of stars on the **main sequence**.

helium flash

The ignition of the nuclear fusion of helium in the center of stars that have left the main sequence. It causes an increased energy output that bloats the star into a **red giant**.

Herbig–Haro object

A "knotty" region of the interstellar medium that emits electromagnetic radiation because it is excited by a jet of gas from the **bipolar outflow** of a protostar.

Hertzsprung–Russell diagram

A graph that plots the absolute magnitudes of stars against their spectral classification. Luminosity can be substituted for absolute magnitudes and effective temperatures for spectral classification. Any star for which these values are known can be included on this diagram. It was developed by both Ejnar Hertzsprung in Europe and Henry Norris Russell in the United States independently of each other.

From left to right the stars follow the **OBAFGKM** pattern (*see* **spectral classification**) and thus go from blue to yellow and to red. The positions of the stars on the diagram are located in specific regions. The **main sequence** is a curving line from the top left of the diagram to the bottom right. It is here that all the stable hydrogen-burning stars are clustered. The Sun is among these and the values on the axes are usually chosen so that it is plotted in the center of the diagram. At the top right are giants and supergiants, at the bottom left are white dwarfs. As stars evolve, they move their position on the diagram. For instance, they drop onto the zero age main sequence along **Hayashi tracks**. As the hydrogen in the core of a star becomes depleted, the star moves across the main sequence. It then leaves the main sequence as hydrogen burning is confined to a shell around an inert core of helium. As that helium ignites, the star swiftly swells to become a red giant. When fusion stops, a low-mass star collapses and becomes a white dwarf; a high-mass star explodes as a supernova.

H II regions

See **emission nebula**.

high-velocity star

A star whose orbit carries it a significant distance out of the plane of the Milky Way. Despite the name, such stars are not actually moving any faster that the other stars around them. However, because they do not share in the general motion of those stars they appear to have much larger relative velocities in unexpected directions.

horizon problem

The problem of explaining how the **cosmic background radiation** became isotropic (the same in all directions) when separate regions of the sky cannot be linked to one another. The cosmic background radiation, from every part of the sky, is a **black body radiation** curve produced by an emitter at 3 degrees kelvin. The problem is that the observable Universe is not yet old enough for light to have traveled from one side of it to the other. Therefore, by implication, thermal reactions to equalize the temperature cannot yet have taken place. Astronomers are therefore puzzled as to how the uniform temperature came into being. The horizon problem and the flatness problem are apparently solved by the cosmological theory of **inflation**.

Hubble constant

See **Hubble flow**.

Hubble flow

The expansion of the Universe brought about by the **Big Bang** which caused space itself to expand, named for the American astronomer Edwin Hubble who measured the expansion of the Universe in 1929. Although celestial objects such as stars and galaxies do not expand, the space in between them does. On a smaller scale than clusters of galaxies, gravity can overcome this expansion. However, the clusters are forever being driven apart by the Hubble flow. The farther apart two clusters are, the greater the rate at which the distance in between them increases. The distance and recessional velocity are linked by a quantity known as the Hubble constant, according to which, for every million parsecs between objects, the space between them grows by 50-100 km every second. This makes it appear as if the objects are receding and causes the electromagnetic radiation they release to be redshifted.

Hubble tuning-fork diagram

A diagram representing the various classes of galaxy, according to the scheme drawn up by American astronomer Edwin Hubble. **Elliptical galaxies** are divided up into eight subclasses from E0, which is a sphere, to E7, which is an oblate spheroid (similar to an American football). These subclasses are placed along the handle of the tuning fork. The two prongs of the fork show three subclasses of **spiral galaxy** and three of **barred-spiral galaxy**. The subclasses define the size of the galactic nucleus and also how tightly the arms are wound. Where the prongs join the handle, an intermediate form of galaxy known as a lenticular galaxy is placed. No irregular galaxies are shown on the tuning fork: they are usually depicted at its side.

hydrogen

The simplest and lightest element in the Universe, in the form of a colorless, odorless gas. Hydrogen makes up 75 percent of the atomic matter present. Its atomic nucleus consists of a single proton. It has two isotopes: deuterium with one neutron, and tritium with two neutrons. It is the nuclear fusion of hydrogen into helium, by the **proton-proton chain**, which generates the energy that stars radiate into space.

hydrostatic equilibrium

The stability reached in some interior regions of stars and planets when the forces acting on the material are in perfect balance. As everything is in equilibrium there is no bulk motion of material.

inflation

See **inflationary cosmology**.

HORIZON PROBLEM

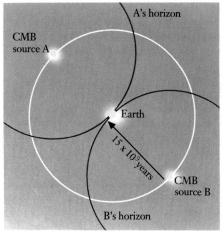

inflationary cosmology

A revision to the standard model of the **Big Bang** to include an epoch when the Universe was 10^{-35} seconds old during which the spacetime continuum expanded faster than the speed of light. Because nothing was traveling through the **spacetime continuum** at a speed greater than that of light, no laws of physics are broken by this theory. The huge inflation of space was driven by the **grand unified force** separating into two: the strong nuclear force and the electroweak force. Tiny isotropic regions (regions the same in all directions) of the early Universe were inflated to become larger than the observable Universe. (*See also* **fundamental forces** and **Planck time.**)

This aspect of the theory provides a satisfactory solution to the **horizon problem** and also to the **flatness problem**. By inflating the Universe so much, any curvature becomes difficult to perceive and so the geometry of the Universe on the largest scale appears flat. Similarly, the surface of the Earth appears flat to us, even though we know it is curved.

CONNECTIONS

NATURAL HISTORY OF THE BIG BANG **78**

INFLATIONARY UNIVERSE **80**

OPEN, FLAT OR CLOSED? **134**

infrared

Electromagnetic radiation with a wavelength longer than visible light. The range is approximately between 10^{-6} and 10^{-4} m. Infrared radiation is often used to study protostars and other objects deeply buried within dense interstellar clouds. Visible light is blocked by these clouds but they can be penetrated by the longer wavelengths of infrared radiation.

infrared astronomy

The study of infrared radiation emitted by celestial objects. By placing a telescope in space, the infrared background can be reduced by a million times. Infrared sources include cool dust clouds and warm dust around young stars in our galaxy, and nearby galaxies, active galaxies and quasars.

initial event

The moment at which the Universe came into existence. *See also* **Big Bang**.

interference

The process by which two waves of electromagnetic radiation (of the same wavelength) combine to form a single wave. When the wave crests coincide, the wave is amplified

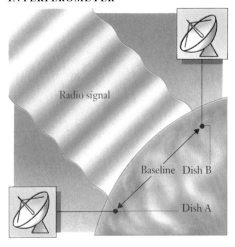

INTERFEROMETER

Radio signal

Baseline Dish B

Dish A

and this occurrence is called constructive interference. When a crest coincides with a trough, the waves cancel out one another and this is called destructive interference.

interferometer

An instrument that increases the ability to separate close objects, using two or more radio telescopes in tandem. Both telescopes observe the same object but from different vantage points. These signals are brought together and an **interference** pattern is formed. From this data an image of the object can be constructed.

intergalactic medium

The matter found in the space between galaxies. In some clusters of galaxies, this space contains very hot, diffuse gas.

CONNECTIONS

CLASSIFICATION OF GALAXIES **88**

CLUSTERS AND VOIDS **94**

interplanetary medium

The diffuse matter found between the planets in the Solar System. Dust is present as fragments from collisions in the asteroid belt and as debris from comets that break up during their passage through the inner Solar System. Charged particles are provided by the Sun in the form of the **solar wind**. There are also neutral gas atoms present which are destroyed close to the Sun but are replenished from the interstellar medium.

interstellar absorption

A process by which interstellar dust and gas block visible light. This is so complete in an **absorption nebula** that no light penetrates at all. Interstellar absorption depends on the wavelength of the electromagnetic radiation

involved: the longer the wavelength, the less the radiation is absorbed.

interstellar cloud

See **giant molecular cloud**.

interstellar medium

The diffuse material to be found between the stars within most types of galaxies. Dust and gas are not distributed uniformly throughout a galaxy, rather they tend to gather into **giant molecular clouds**. Within these clouds are dense cores and glowing **emission nebulas**. Density wave theory describes the interaction of the arms of a spiral galaxy with the interstellar medium. It is believed that elliptical galaxies contain very little dust and gas.

CONNECTIONS

ACTIVE GALAXIES **96**

ENERGY MACHINE **98**

LIFE AND DEATH OF A STAR **116**

ion

An atom that has acquired an electrical charge through **ionization**. Positive ions have an incomplete quota of electrons as a result of this process. The free electrons are the negative ions.

ionization

A process during which electrons gain sufficient energy to escape completely from an atom. The required energy is provided by collisions with other particles or **photons**.

IRAS

The InfraRed Astronomical Satellite. This **infrared** telescope was launched during 1983 and observed the Universe at infrared wavelengths, in unprecedented detail, for most of that year. It was a **reflecting telescope** with mirrors made of beryllium, rather than glass, to withstand the low temperatures at which it was designed to operate (2 kelvin). During the 10 months in which it was operational, it mapped 96 percent of the sky and detected a quarter of a million individual infrared sources, some of which were previously unknown objects.

iron

A metallic element, with 26 protons in its nucleus. Iron is formed in the core of the most massive red giants, and is the heaviest element thus formed.

irregular galaxy

A galaxy that does not fit into the classifications on the **Hubble tuning fork diagram** because its shape appears random and un-ordered. When studied carefully, such a galaxy may exhibit a disklike motion and structure. If so, it is called a type I irregular. If not, the galaxy is a type II irregular. In some irregular galaxies star formation can be seen taking place. Up to one quarter of all galaxies are irregular.

isotopes
One of two (or more) atoms of the same element that differ from each other only in the number of neutrons they contain within their atomic nucleus.

IUE
The International Ultraviolet Explorer. An orbiting space telescope, launched from the Earth in 1978. It was a traditionally designed reflecting telescope which worked at **ultra-violet** wavelengths. It functioned successfully for 13 years.

Jupiter
The largest planet in the Solar System, a gas giant planet with a diameter 11 times that of the Earth. The fifth planet in order from the Sun, Jupiter marks the beginning of the outer Solar System. Although the most massive of the planets, it is still dwarfed by the Sun and has only 0.1 percent of a solar mass. Its composition is similar to the Sun's, however – it is made up mostly of hydrogen and helium with only traces of the other elements. Jupiter, in common with the other gas giants, has no solid surface. Instead, its atmosphere becomes thicker and thicker until the hydrogen is compressed into exotic states of matter which behave more like a metal than a gas. Jupiter radiates twice as much heat as it receives from the Sun, and its poles are as warm as its equator. The cloud tops contain belts and zones of color, thought to indicate the depth and composition of the clouds. Also visible are huge cyclonic storms such as the Great Red Spot. Jupiter possesses 16 natural satellites, four of them easily visible to observers on Earth. They are known as the Galilean satellites and, in order of distance from Jupiter, are Io, Europa, Ganymede and Callisto.

Kaluza–Klein theory
A theory that tries to unify the strong nuclear force, the weak nuclear force, the electromagnetic force and gravity. It postulates that the spacetime continuum consists of more than 10 dimensions of which we perceive only three (up and down, left and right, in and out) because the others are too tightly wound, but that we feel the effects of these others by perceiving them as the fundamental forces.

KEPLER'S LAWS

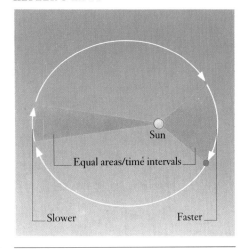

Kelvin
A temperature scale based on a zero point at which atoms in solids stop vibrating (absolute zero). Zero kelvin corresponds to –273.16° C.

Kepler's laws of planetary motion
The three laws that explain the motion of objects in elliptical orbits, developed by Johannes Kepler and announced by him in 1609 and 1618. They apply to any object in circular or elliptical orbit around the Sun and equally to satellites, both natural and artificial, in orbit around a planet. The first law states that planets move in elliptical orbits with the Sun at one focus; the second that equal areas are swept out by the orbits in equal times, so the planets must be traveling faster when they are closer to the Sun. Law three holds that the square of the planet's orbital period is proportional to the cube of its mean orbital radius from the Sun.

CONNECTIONS

A UNIVERSE OF RULES **48**

FORCES AND FIELDS **68**

THE SUN **106**

Kuiper belt
A flared disk of icy bodies that has its inner boundary just beyond the orbit of the planet Pluto. The Kuiper belt supplies the Solar System with short-period comets. With increasing distance from the Sun it extends gradually upward and around into the Oort cloud that surrounds the Solar System.

Lagrangian point
A position in the orbital plane of a two-body system at which a much less massive object can remain in equilibrium. There are five theoretical Lagrangian points in such a sys-

tem but only two of these are stable, resisting perturbations by outside gravitational forces. They occur on the orbit of the smaller of the two massive bodies, 60° from them both, at one point of an equilateral triangle.

last scattering surface
See **decoupling of matter and energy**.

laws of physics
Mathematical equations and rules that predict the behavior of the Universe. They refer to quantities that can be observed and measured. The laws of physics, which include physical constants such as the speed of light and the forces of nature, are thought to have been shaped during the few instants, known as the Planck time, that followed the **Big Bang**.

length contraction
A relativistic effect that is observed from an object traveling at speeds aproaching the speed of light. It was predicted by Einstein's theory of special relativity. From a spacecraft traveling at close to the speed of light, the lengths of objects in the outside Universe appear to contract. The length of the spacecraft also appears to contract when observed from the outside Universe. *See also* **time dilation** and **mass dilation**.

lenticular galaxy
An intermediate form of galaxy, between elliptical galaxies and spiral galaxies. Lenticular galaxies possess flattened forms and galactic disks but have no spiral arms.

Leo
The Lion, a constellation on the ecliptic and fifth sign of the **zodiac**. It contains the bright star Regulus (apparent magnitude +1.35). Leo gives its name to the annual Leonid meteor shower.

LAGRANGIAN POINT

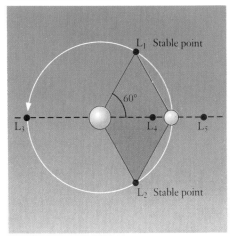

lepton

An elementary particle such as an electron or neutrino which does not take part in interactions involving the **strong nuclear force** and is not composed of quarks.

Libra

The Scales, an average-size constellation on the ecliptic; the seventh sign of the **zodiac**.

light curve

A graph showing the changes in magnitude of a celestial object plotted against time. Brightness is usually measured by **photometry**. The usual objects studied are **variable stars** and **eclipsing binaries**.

light-year

A unit of astronomical measurement, the distance traveled by electromagnetic radiation in one year. It is equivalent to 9.46×10^{12} km. *See* **parsec**.

line spectrum

A spectrum consisting solely of **spectral emission lines**. Such spectra are observed from luminescent gas clouds in which electrons yield energy in the form of photons.

Local Group

The group of a few dozen galaxies of which our own Galaxy is a member. There are three spiral galaxies: the Milky Way, Andromeda and M33 – and a giant elliptical galaxy, Maffei I, as well as small ellipticals and irregular galaxies. The shape of the Local Group is like a dog bone with concentrations of galaxies around the Milky Way and Andromeda. The radius of the Local Group is approximately 3.25 million light-years.

Local Supercluster

The supercluster of galaxies, centered on the Virgo cluster, to which the Local Group belongs. It measures about a hundred million light-years across.

Lorentz transformation

A set of equations used in relativity problems to transform measurements from one frame of reference to another.

luminosity

The total energy radiated into space every second by a celestial object such as a star.

luminosity class

The classification of a star according to its absolute magnitude and spectral type. Luminosity classes help to pinpoint the position of a star on the **Hertzsprung–Russell diagram**. Class I stars are supergiants, class II stars are bright giants, class III stars are giants, class IV are subgiants and class V are dwarfs (main sequence stars).

Magellanic cloud

One of two satellite galaxies of the Milky Way, the Large and the Small Magellanic clouds. Both are irregular galaxies in which stars are forming. They are visible only from the Southern Hemisphere and are 60,000 parsecs away (1 parsec = 32.6 light-years). Their diameters are 7000 and 2500 parsecs.

magnitude

A measure of the brightness of a celestial object. **Apparent magnitude** is a measure of how bright the object appears, without taking into account its distance away. **Absolute magnitude** provides a standard of comparison between objects of varying distance.

main sequence

A region on the **Hertzsprung–Russell diagram** in which stable, middle-aged stars are found. These stars are burning hydrogen in their cores. When the hydrogen ignites, the stars pass through the **T Tauri phase** and arrive at the **zero-age main sequence**. As the stars age and use up the hydrogen, they change their luminosity and their spectral classification and move across the main sequence. When the hydrogen burning in the core of a star eventually gives way to helium fusion, the star moves right off the main sequence to become a **red giant** star.

Mars

The fourth planet in distance from the Sun. It orbits in a distinct ellipse but has a mean distance from the Sun of 1.52 astronomical units (1AU = 149.6 million km). It is the outermost of the four terrestrial planets. Beyond Mars lies the asteroid belt and the gas giants of the outer Solar System. An estimated one quarter of the mass of the planet consists of an iron core. The rest overlies it as rocky material extending to the crust. Mars has a tenuous atmosphere of carbon dioxide gas. It freezes onto the polar regions during the Martian winter and sublimes back into a gas during the summer months when the planet moves closer to the Sun again. Mars has two moons, Phobos and Deimos.

mass

A fundamental property of most particles of matter. The amount of mass contained in an object contributes to the strength of its gravitational field, as does the object's density. Some particles, such as **photons**, carry energy but have no mass when at rest.

mass defect

The quantity of mass converted into energy during nuclear fusion or nuclear fission. The conversion takes place according to the equation $E = mc^2$, where E is energy, m is the mass defect and c is the speed of light.

mass dilation

The increase in mass that an object undergoes when it approaches the speed of light. It is a consequence of special relativity. *See also* **length contraction** and **time dilation**.

matter

Any particle that possesses mass, for example a proton or an electron. Particles that just carry energy such as photons and neutrinos do not constitute matter.

megaparsec

A unit of astronomical measurement, one million **parsecs** (1 parsec = 3.26 light-years).

Mercury

The closest planet to the Sun and the smallest of the terrestrial planets. It is somewhat similar in appearance to Earth's only natural satellite, the Moon, in that it is heavily cratered. It has a substantial core of ferrous metals, which is thought to account for 70 percent of the mass of the planet and three-quarters of the planet's volume. Mercury does not have an atmosphere, however, because it is too small to possess the gravity necessary to hold gases to its surface. It has no natural satellites.

meridian

The great circle on the **celestial sphere** that passes through both poles and the observer's zenith.

meson

A particle composed of two quarks. More precisely, a meson is made up of a **quark** and its **antimatter** counterpart, bonded together. Mesons exist only for a very short time because the quark and antiquark soon annihilate one another. *See also* **elementary particle** and **fundamental particle**.

Messier catalogue

A catalog of bright galaxies, nebulas and star clusters published by the French astronomer Charles Messier in 1774. The original list contained only 45 objects, but it was added to in the centuries that followed. It now contains 110 objects, although some errors have crept in. For example, M102 is a duplication of M101. The catalog is still in wide use, however. For instance, M31 is the Andromeda galaxy.

metals

In astronomy, the generic term for all of the elements heavier than helium. Astronomers group this diverse group of chemicals together because they contribute no more than 2 percent of the atomic matter in the Universe. Seventy-five percent is hydrogen and the other 23 percent is helium.

microwaves

Electromagnetic radiation with a wavelength between 1 and 100 mm. This region of the electromagnetic spectrum is bounded by **radio waves** and **infrared** radiation.

Milky Way

The name of our own Galaxy, which contains about one hundred billion stars. At present, the Sun is in the trailing edge of the Orion spiral arm, about two-thirds of the way toward the edge of the Milky Way. All the stars visible in the night sky are in our own galaxy. The misty band of light that stretches across the night sky and is traditionally called the Milky Way is our view of the galactic disk. The Milky Way has a diameter of 100,000 light-years and may be a **barred-spiral galaxy**. It possesses a galactic halo of globular clusters of stars and is held by gravity to a few dozen other galaxies, known as the **Local Group**.

CONNECTIONS

SCALE OF THE UNIVERSE 76

THE MILKY WAY 92

Mira-type variable

Red giant and red supergiant star that varies the amount of energy it emits over periods of several months. Because these variations are repeated at regular intervals, it is believed that Mira-type variables are due to the stars suffering pulsation in their surface layers.

moon

A naturally occurring, relatively large body in orbit around a planet.

Moon, the

The Earth's only natural satellite. It is a heavily cratered body whose composition is deficient in the heavier, metallic materials. This is thought to be due to the way that the

Moon was formed, from some of the material thrown out when a massive body collided with the forming Earth. The heavier ejecta did not reach Earth's orbit: it fell back to the planet under the pull of gravity. The lighter material escaped and eventually formed the Moon. The Moon has no permanent atmosphere or running water, so it has never suffered erosion or weathering. Thus, it presents scientists with a cratering record that dates back to the forming of the Solar System. Because of this, it provides invaluable information for understanding how the planets came into being.

muon

A **lepton** with a negative charge, with a mass 206 times that of an electron. *See also* **elementary particle**.

nadir

The point on the **celestial sphere** directly below the observer, opposite to the zenith.

NASA

The National Aeronautics and Space Administration – the United States government agency set up in 1958, responsible for designing and implementing the US presence in space. NASA was responsible for the Apollo Moon landings, for the Viking probes which reached Mars during the mid-1970s and for the Voyager fly-bys of the outer Solar System planets in the 1980s. NASA administers and operates the Space Shuttle and is looking into exploration of the Solar System's planets.

nebula

A cloud of dust and gas in space that is visible to observers on the Earth because it either emits, reflects or absorbs starlight. *See also* **absorption nebula**, **emission nebula** and **reflection nebula**.

Neptune

The eighth planet from the Sun in the Solar System. Neptune is part of the outer Solar System and is a **gas giant** planet. During the years 1979 to 1999, it is actually the outer most planet in the Solar System, because Pluto has such an eccentric orbit that it crosses over inside Neptune's orbit during those years. Neptune is a dynamic world with cloud features and cyclonic storms, somewhat similar to those of Jupiter and Saturn. However, Neptune is composed of many more icy substances such as methane and ammonia than Jupiter or Saturn. The methane in Neptune's atmosphere causes it to appear blue. Neptune has eight moons, most of which are small rocky bodies. One of them, Triton, is an interesting world in its

own right. It has a complex set of seasons and a experiences a strange icy form of volcanic activity on its surface.

neutrino

An elementary particle possessed with negligible or no mass and no charge. Research is taking place into whether neutrinos have a small mass but, at present, it is assumed that they do not. Neutrinos are produced in reactions involving the weak nuclear force and they carry away excess energy from the reaction. They hardly react at all with baryon particles and have been postulated to be dark matter. If they are the **dark matter** content of the Universe, it will need to be proved that they have mass.

CONNECTIONS

THE PARTICLE FAMILY 62

NATURAL HISTORY OF THE BIG BANG 78

COLLAPSING AND EXPLODING STARS 132

---

neutrino decoupling

An event that is supposed to have taken place when the Universe was one second old, by which time the density of matter in the expanding Universe had fallen far enough for neutrinos no longer to be subjected to constant interactions with other particles.

neutron

An **elementary particle** with zero charge. It is composed of three quarks and is a constituent of every atomic nucleus except normal hydrogen. Its mass is slightly more than that of a proton, and if released from an atomic nucleus, a neutron decays, because of the **weak nuclear force**, into a **proton**, an electron and an antimatter neutrino.

neutron star

The central remains of a star that has undergone a **supernova** explosion. It is composed of baryon degenerate matter and has a very high density. Typically, its radius is just 10 km whereas its density may be as much as 10^{18} kg/m^3. It was once thought that neutron stars were impossible to detect by observation, but it seems that **pulsars** are spinning neutron stars. Moreover, X-ray bursters are now thought to be binary star systems in which one of the components is a neutron star. The mechanism for outburst is similar to that in a **nova**.

CONNECTIONS

KINDS OF STARS 102

NEUTRON STARS AND PULSARS 128

New General Catalogue

A catalogue of nonstellar celestial objects that was compiled by J.L.E. Dreyer of Armagh Observatory, Ireland, and published in 1888. It listed 7840 objects and was supplemented by a further list that now includes another 5386 objects. Those on the original list are identified by the letter NGC followed by the catalog number. Objects on the supplement are denoted by the letters IC followed by the catalog number.

nova

A star that is observed to brighten suddenly by up to 10 magnitudes and then gradually decline back to its original magnitude over a period of months. However, careful study has revealed that, rather than being single stars, novae are close binary star systems in which one of the members is a white dwarf star. Its companion has evolved and now fills the **Roche lobe**, thus allowing matter to be transfered from it to the white dwarf via the inner Lagrangian point. This material forms an **accretion disk** around the white dwarf and gradually spirals down onto its surface. There the matter, usually hydrogen atoms, builds up until the temperature and pressure are great enough to ignite nuclear fusion through the **carbon cycle**. This causes a runaway chain reaction and a nuclear detonation. The star system drastically increases in brightness as the energy is liberated.

nuclear fission

A process by which an atomic nucleus splits into lighter elements. This is the predominant nuclear reaction for heavy elements, which give out energy when it occurs.

nuclear fusion

A process by which two or more atomic nuclei join together to make a heavier one. This is the type of nuclear reaction that takes

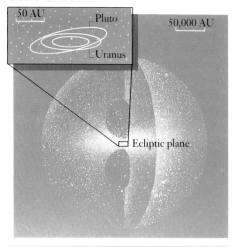

OORT CLOUD

50 AU

Pluto

50,000 AU

Uranus

Ecliptic plane

---

place in the center of stars. The lighter elements give out energy in the process of fusing. *See* **proton–proton chain**.

nucleosynthesis

The process by which the chemical elements are built by naturally occurring nuclear reactions. Such a period took place when the Universe was between one and five minutes old. It now occurs in the center of stars and during supernova explosions. It can also be found in high-energy collisions between atomic nuclei and high-energy elementary particles such as those found in cosmic rays.

OBAFGKM

See **spectral classification**.

OB association

A region of space in which recent star formation has led to the birth of high-mass O and B type stars. There can be a couple or a couple of hundred such stars in each association. The intense ultraviolet radiation produced by these stars ionizes and excites the sur-

rounding hydrogen, causing **emission nebulas,** or H II regions, to be produced. The vast stellar winds given out by these stars sweep out a cavity into which the H II region expands. A shock wave forms which compresses the dust and gas of the interstellar medium, leading to gravitational collapse and the formation of more stars.

CONNECTIONS

STARS AND GALAXIES **104**

ON THE MAIN SEQUENCE **122**

Oort cloud

A spherical region, believed to surround the Solar System, which contains a vast number of comets. The comets are disturbed by the passage of nearby stars and fall in toward the inner Solar System. Comets coming from the Oort cloud are known as long-period comets because they may have orbital periods measured in thousands of years. Some have such large eccentric orbits that they will eject them from the Solar System altogether. The Oort cloud is thought to exist between the distances of 30,000 to 100,000 astronomical units (1AU = 149.6 million km). As yet there is no positive proof of its existence.

open universe

Any model of the universe in which the ratio of its mean density to its critical density is less than one. In that case the universe does not contain enough matter to halt its expansion through the action of gravity. Hence it will expand forever, with its contents gradually becoming more and more scattered throughout the cosmos. However, on this model, at some point in the far future, stars, planets and galaxies will no longer exist because the universe will have reached thermal equilibrium. This is known as the **heat death** and is the ultimate fate of an open universe, or of a "flat" universe.

Oppenheimer–Volkoff limit

The mass limit at which **baryon degeneracy pressure** is no longer sufficient to resist gravity. This occurs for inert masses of between three and five solar masses. It is the upper limit for the mass of a **neutron star** and the lower limit for that of a **black hole**.

optical binary

A pair of stars that lie close to one another on the **celestial sphere** because of a chance alignment. They are not physically associated with one another and probably exist at vastly different distances. Optical binaries are also known as visual binaries. *See also* **binary star**.

OB ASSOCIATION

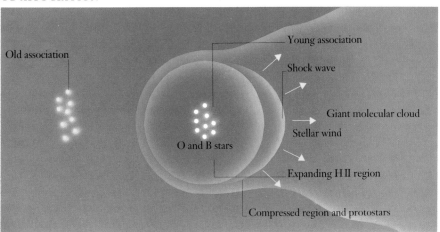

Old association

Young association

Shock wave

Giant molecular cloud

Stellar wind

O and B stars

Expanding H II region

Compressed region and protostars

optical depth

A measure of how difficult it is for a **photon** of electromagnetic radiation to pass through an optical medium. A dusty medium has a high optical depth (it is "optically thick") because a photon stands a high probability of interacting with a dust particle. This prevents the photon from continuing along its path by either scattering it or absorbing it. When this phenomenon occurs within the interstellar medium, it is known as **interstellar absorption**. Low optical depths ("optically thin") mean that photons can pass through the medium relatively unhindered.

optical spectrum

The part of the full electromagnetic spectrum that is confined to the wavelengths of visible light – between 400 and 760 nm.

orbit

The path that an object takes under the influence of gravity. Orbits follow the shape of conic sections, and may therefore be circular, or elliptical (as in a comet in orbit around the Sun). Celestial objects of similar masses may orbit each other around a common center of gravity, especially binary stars. *See also* **escape velocity**.

Orion

An imposing constellation that is visible to observers in both hemispheres of the Earth. Orion is familiarly called the Hunter and contains many objects of interest, including **emission nebulas**, **absorption nebulas**, an **OB association**, a **giant molecular cloud** and a **red giant** star.

pair production

The formation of an electron and its anti-matter counterpart, a positron, from a gamma ray. Pair production occurs when a gamma ray passes close to an atomic nucleus.

ORBIT

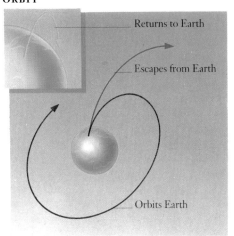

Returns to Earth

Escapes from Earth

Orbits Earth

parabola

An open curve and one of the **conic sections**. A three-dimensional parabola is known as a paraboloid. The shape is used in a reflecting telescope for the primary mirror.

parallax

The apparent angular displacement of an object, caused by viewing it from slightly different positions. Parallax is used in astronomy to determine the distance to nearby stars. A star is observed from Earth once, then observed again six months later. These two readings provide the maximum possible difference in position, that is two **astronomical units**. The angular displacement of the star can then be used to calculate its distance by trigonometry.

parsec

A unit of astronomical measurement, defined as the distance at which a star would have a parallax of one arcsecond when viewed from two points the distance apart of the Earth and the Sun. It is equivalent to 3.26 light-years.

particle

A single piece of matter (for example, proton, electron, pi meson) or an energy carrier (for example, photon, neutrino).

particle accelerator

A device, usually very large, that uses magnetic fields in accelerating particles to speeds close to the speed of light. The particles are forced to collide with other particles and the results are recorded using devices such as **bubble chambers**. Particle accelerators may be circular, as in the European accelerator at CERN, on the French–Swiss border, or linear, as in the LINAC at Stanford, California, in the United States. The results help scientists to understand high-energy physics and the conditions that were present in the early Universe.

Pauli's exclusion principle

A fundamental principle of quantum theory which states that no two **fermions** in a single system can exist in the same **quantum state**. The principle describes the physical processes that lead to the structure of atoms. It also helps to describe the conditions under which degenerate matter exists. It therefore explains how **white dwarfs** and **neutron stars** support themselves against the downward pull of gravity.

peculiar velocity

The velocity of an individual star relative to the average velocity of the stars in the solar neighborhood. It is also used to describe the

velocity of individual galaxies apart from that which is due to the **Hubble flow**.

periastron

The point in the orbit of an object orbiting a star, at which the object is closest to the star.

perigee

The closest point to the Earth that is reached by an object orbiting the Earth.

perihelion

The point in its orbit in which a member of the Solar System is at its closest to the Sun.

Periodic Table

A table in which all the chemical elements are grouped in order of mass. By arranging the elements in rows and columns, periodic similarities in chemical and physical properties can be seen. It has been important in the discovery of new elements and the understanding of atomic structure and chemical reactions.

perturbation

The disturbance in motion produced by one celestial object on another.

phase change

A change in the way that atoms, molecules or particles aggregate in a substance. The different aggregations are known as phases. For instance, when water exists as a gas in clouds, changes in temperature and pressure can cause it to change its phase: the water molecules aggregate into a liquid and the water falls as rain. A further phase change occurs when the water solidifies to form ice.

photometer

A device used in photometry to measure the brightness of celestial objects. Colored filters are used to measure the brightness of an object at different wavelengths.

photometry

The accurate measurement of the magnitudes of stars and other celestial objects, using instruments such as a photometer.

photon

An **elementary particle**, probably of zero mass, that carries energy generated in reactions involving electromagnetism. The photon is the particle associated with electromagnetic radiation according to **wave-particle duality**. The type of electromagnetic radiation is determined by the energy of the photon. For example, the highest energy photons are gamma rays and the lowest energy photons are radio waves. *See also* **fundamental particle**.

photosphere
The surface of any star, for example the Sun. It is a layer that emits visible light and in the Sun is about 500 km deep. The conditions cause gas in this region to alter its **optical depth** from thick to thin and so it becomes transparent to visible light and can be seen. The temperature of the Sun's photosphere is approximately 6000 kelvin, which drops to about 4000 kelvin at the base of the **chromosphere**. The gas visible on the photosphere is in constant motion. This motion is deep convection and produces the pattern known as granulation on the photosphere. **Sunspots** occur in the photosphere.

Pisces
The Fishes, a large, faint constellation on the ecliptic, the 12th sign of the **zodiac**. The vernal equinox now lies in Pisces.

Planck constant
The ratio of a particle's energy to its frequency, named for the German physicist Max Planck who proposed it in 1900. Its value is always equal to 6.626×10^{-34} joule seconds. It governs the accuracy with which different properties can be measured simultaneously (*see* **Heisenberg's uncertainty principle**), and the wavelength associated with a particle.

Planck time
The interval of time within which the laws of physics break down. It is equal to 10^{-43} seconds and is important in the **Big Bang** theory. When the Universe was less than 10^{-43} seconds old, gravity is thought to have been unified with the **grand unified force**.

CONNECTIONS

UNIFYING THE FORCES **70**

NATURAL HISTORY OF THE BIG BANG **78**

INFANT UNIVERSE **82**

---

planet
A rocky, gaseous or icy body in orbit around our Sun or any other star. In our own Solar System there are nine planets: Mercury, Venus, Earth, Mars, Jupiter, Saturn, Uranus, Neptune and Pluto.

planetary nebula
The outer, gaseous layers of a **red giant** star that have been blown off into space, while the stellar core is in the process of becoming a **white dwarf**. The name derives from 18th-century British astronomer William Herschel's belief that they resemble planets under low magnification through telescopes. Planetary nebulas have been discovered that are neither ring- nor shell-like but bipolar and irregular. They are all expanding and have estimated lifetimes of less than 40,000 years, after which they become too tenuous to be seen. They shine because the gas in them is excited by **electromagnetic radiation** from the central, collapsing star.

plasma
The fourth state of matter (after solid, liquid and gas), which occurs when a gas is heated until every atom and molecule becomes ionized (charged). The plasma exists as independent positive and negative **ions**.

Pluto
The outermost planet in the Solar System. It is also the smallest planet and is chiefly composed of icy material such as frozen methane. Its highly elliptical orbit brings it inside the orbit of Neptune during some of its "year". A Plutonian year is equal to 248 Earth years. Pluto has one moon, named Charon. It, too, is an icy world.

Polaris
Alpha Ursae Minoris, a Cepheid variable star in the Little Dipper constellation. It is known as the Pole Star and has been used for navigation for many centuries although, owing to precession, it is moving slowly away from the north celestial pole.

Population I stars
A grouping of stars made by German-born astronomer Walter Baade in about 1944, indicating the younger stars, which are relatively rich in metals. They are found in the arms of spiral galaxies, associated with clouds of interstellar matter. The Sun is a Population I star.

Population II stars
A grouping of stars made by German-born astronomer Walter Baade in about 1944, indicating older stars, found mainly in elliptical galaxies, the centers of spiral galaxies, between the spiral arms and in globular clusters. They are deficient in metals.

positron
The **antimatter** counterpart of an electron.

post-main sequence
The phase of a star signified by the ignition of helium in its core. It is the position a star holds following its evolution off the main sequence on the **Hertzsprung-Russell diagram**, on its way to becoming a red giant.

potential energy
The energy contained within an object which is suspended in a gravitational field. It can be released by the object falling under the influence of the gravitational field.

potential well
The distortion caused to the **spacetime continuum** by the presence of an object with mass.

precession
The apparent slow movement of the celestial poles on the celestial sphere. It is caused because the Earths rotation axis rotates around a fixed position in a period of 25,800 years.

pre-main sequence
The stage in the life of a star that has passed the protostar stage but has not yet begun stable nuclear fusion of hydrogen in its core and has not yet reached the main sequence on the **Hertzsprung–Russell diagram**.

CONNECTIONS

BIRTH OF A STAR **118**

FORMATION OF PLANETS **120**

---

principle of equivalence
A principle of **general relativity** that states that an observer in a closed system cannot tell whether the force acting upon him or her is gravity or whether the entire system, in which the observer is placed, is being accelerated. In other words: the effect of gravity can be counteracted by applying equal but opposite acceleration. By regarding gravity as an accelerated frame of reference, general relativity can predict its effects.

principle of relativity
The principle that every measurement taken in the Universe has to be relative to some frame of reference, usually that of the observer. It is central to both **special relativity** and **general relativity**.

prominence
A cloud or plume of hot, luminous gas that is elevated into the lower solar **corona** from the top of the **chromosphere**. This usually occurs over regions of solar activity such as **sunspot** groups. Prominences are supported by magnetic fields and are of several kinds. In a surge prominence, gas moves up and down along the same path. In a loop prominence, the gas is suspended in a loop. Arch prominences resemble hedgerows of glowing gas. Some prominences exist briefly, whereas others are quiescent and last for months, almost unchangingly. They are best observed at the limb of the Sun. When they are in silhouette against the disk they are known as filaments.

protogalaxy

A galaxy that is in the process of formation. As yet, no conclusive observations of one have been made.

proton

An elementary particle composed of three quarks and carrying a positive electrical charge. It is a constituent of all atomic nuclei. The number of protons in a nucleus of an atom determines the chemical element.

> **CONNECTIONS**
>
> INSIDE THE ATOM **50**
>
> PARTICLE EXPERIMENTS **52**
>
> SCALE OF THE UNIVERSE **76**
>
> NATURAL HISTORY OF THE BIG BANG **78**

proton-proton chain

The sequence of nuclear fusion reactions in which a helium nucleus is built up by the successive collisions of hydrogen nuclei. Hydrogen nuclei are single protons, hence the name. The proton-proton chain releases energy and it is this reaction that is thought to power most **main sequence** stars.

protostar

A star in the process of formation. Detailed observations are continually being made of these objects by astronomers. They are found in the dense cores of **giant molecular clouds**. Because they are surrounded by dusty envelopes, protostars must be studied with long wavelengths of electromagnetic radiation, such as infrared emission, because that is not attenuated as much as visible light. Protostars have no nuclear reactions taking place within them.

> **CONNECTIONS**
>
> BIRTH OF A STAR **118**
>
> FORMATION OF PLANETS **120**

pulsar

A spinning **neutron star**. A pulsar is visible because of two beams of electromagnetic radiation that emanate from the magnetic poles and sweep across our line of sight. There are two types of pulsars. One type spins relatively slowly: for example, the pulsar in the Crab Nebula supernova remnant. This type slows down as time passes. The second type is found in **binary star** systems and has been made to spin faster by the transfer of material from the companion star. This type is known as a millisecond pulsar. Both types are created from the remains of a star that has exploded as a **supernova**.

quantum chromodynamics

A theory that explains the strong nuclear force in terms of hadrons being composed of quarks. The quarks communicate by exchanging massless **virtual particles** known as gluons.

quantum state

A fixed set of quantum properties, such as energy, angular momentum and spin, with which particles can exist. For example, electrons can exist in many states around atoms, but no two electrons exist in the same quantum state (**Pauli's exclusion principle**). This means that atoms have a definite structure and cannot normally be compressed.

quantum theory

A theory whose guiding principle is that physical quantities are not continuous but can have only discrete values. It was developed in the early 20th century by many contributing scientists, including the German physicists Erwin Schrödinger and Werner Heisenberg and the Dane Niels Bohr. It states that nature restricts the values that properties can take; this is known as quantization. In the everyday world, the size of this quantization is so small that it goes unnoticed. Only on the scale of subatomic particles is it noticeable.

> **CONNECTIONS**
>
> THE QUANTUM VIEW **64**
>
> EXCLUSION AND UNCERTAINTY **66**
>
> UNIFYING THE FORCES **70**

quark

A fundamental particle that joins with others in triplets and doublets to form hadrons such as protons and pi mesons. In the present-day Universe, quarks must exist as triplets or doublets. In the early Universe they existed as free particles. Mesons are not charged and are composed of a quark and an antiquark. Quarks are held together by the interquark force that provides the foundation for the **strong nuclear force**.

> **CONNECTIONS**
>
> PARTICLE EXPERIMENTS **52**
>
> THE PARTICLE FAMILY **62**
>
> SCALE OF THE UNIVERSE **76**
>
> NATURAL HISTORY OF THE BIG BANG **78**
>
> INFANT UNIVERSE **82**

quasar

A compact extragalactic object that is far away and highly luminous. Quasars are thought to be **active galactic nuclei**. The name comes from "quasi-stellar radio source". A quasar is very similar to a quasi-stellar object (QSO) but it also gives out radio waves. Thousands have been observed, all at extreme distances. This has led many astronomers to think that they may be young galaxies. They are just one type of the many **active galaxies** now visible.

> **CONNECTIONS**
>
> ACTIVE GALAXIES **96**
>
> ENERGY MACHINE **98**

radiation pressure

The force exerted by a **photon** of electromagnetic radiation when it collides with a particle of matter. The greater the energy of the photon, the greater the radiation pressure that is exerted.

radiative zone

The region in a star where the primary form of energy transportation is by means of constant absorption and re-emission of photons, the quanta of electromagnetic radiation.

radioactive decay

The spontaneous disintegration of certain atomic nuclei by emission of either an atomic nucleus of helium (alpha decay) or an electron (beta decay). It is sometimes accompanied by the emission of gamma rays and causes a change of the atomic nucleus into another, usually more stable element. On production of a stable nucleus, the radioactive decay stops. The time taken for half the atoms in a sample of an isotope to undergo radioactive decay is known as its **half-life**.

radioactivity

See **radioactive decay**.

RADIOACTIVE DECAY

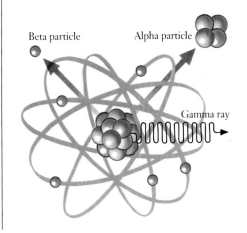

Beta particle　　Alpha particle

Gamma ray

radio galaxy

An elliptical galaxy that is an intense source of **radio waves**. It is a form of **active galaxy**. The radio emission usually comes from two huge lobes on each side of the galaxy. The lobes can be millions of light-years across and are often credited with being the largest single objects in the entire Universe. The radio waves are produced by the **synchrotron emission** process.

radio telescope

A telescope designed to observe **radio waves**, usually comprising a large dish focusing the radio waves onto a collector. Two or more radio telescopes can be linked together to form an **interferometer**.

radio wave

A ray of electromagnetic radiation whose wavelength exceeds 1 m. It is the least energetic form of electromagnetic radiation.

recurrent nova

A **nova** that has been observed to erupt in brightness more than once.

red dwarf

Any star on the main sequence of the **Hertzsprung–Russell diagram** that has a spectral classification of K or M.

red giant

Any star burning helium in its core that has a **spectral classification** of K or M. It is enlarged and appears red because of the change it has undergone in its surface temperature.

redshift

The lengthening of the wavelength of **spectral lines**, caused either by the motion of the source away from the observer, or by the motion of the observer away from the source. Spectral lines from distant galaxies are redshifted. This is interpreted to mean that the galaxies are moving away from Earth, due to the expansion of the Universe. Redshift is therefore used to calculate the distance of extremely far-away objects such as quasars. *See also* **Doppler effect**.

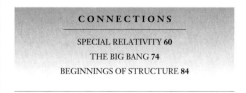

red supergiant

An evolved star that has an even larger radius than a **red giant**.

reflecting telescope

A telescope that uses mirrors to focus the light. There are several types. A Newtonian telescope uses a parabolic primary mirror and a flat secondary mirror, whereas a Cassegrain uses a parabolic primary and a convex hyperbolic secondary mirror.

reflection nebula

A type of interstellar cloud that scatters starlight into our line of sight. Such nebulas often appear blue because blue light, having a smaller wavelength, scatters more efficiently than red light.

refracting telescope

A telescope that uses lenses to focus the light from distant celestial objects. Because a large-aperture telescope would require a very long tube, refractors are usually small.

relativistic speed

Any speed that is close to the speed of light. At these speeds, relativistic effects such as **time dilation**, **length contraction** and **mass dilation** take place.

rest mass

The mass of a particle when it is not moving. The faster an object travels the more the mass increases because of the **mass dilation** effect. This property is imperceptible until

REDSHIFT

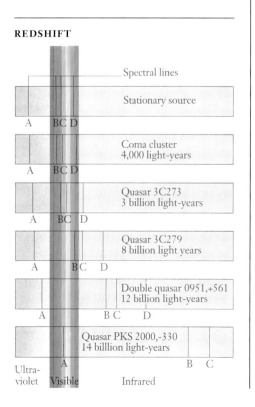

relativistic speeds, approaching the speed of light, are achieved. Some subatomic particles do move at such speeds and hence they have increased masses. To distinguish between a stationary particle's mass and the variable one that it acquires due to speed, rest mass is always quoted.

right ascension

One of the ordinates referring to objects on the **celestial sphere**. It is the equivalent to a longitude reference on the Earth. There are 24 hours of right ascension within 360°, so one hour is equivalent to 15°. Together with **declination**, it represents the most commonly used coordinate system in modern astronomy. Zero hours is the point at which the **ecliptic** crosses the projection of the Earth's equator onto the celestial sphere.

ring galaxy

A rare type of galaxy that is thought to come into being when a compact galaxy passes right through the center of a normal spiral galaxy. This encounter causes a shock wave, which passes outward in a concentric circle and triggers star formation by compressing the interstellar medium in its path. These newly formed stars shine in a ring.

Roche lobe

A pear-shaped region around the stars in a **binary star** system. The lobes define the space within which each star's gravity dominates. They touch at the inner **Lagrangian point**.

ROSAT

An X-ray telescope, currently in orbit around the Earth. It was built by Germany with support from NASA and the United Kingdom. It has made a map of the entire sky at X-ray wavelengths and also in the extreme ultraviolet wavelengths.

ROCHE LOBES

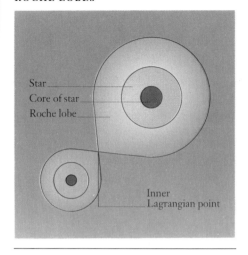

Star
Core of star
Roche lobe
Inner
Lagrangian point

RR Lyrae variable star

A variable star found in galactic nuclei and globular clusters. RR Lyrae variable stars are giant stars of stellar Population II, and have **spectral classifications** between A and F. They typically have periods of only a few hours. Their upper period limits are just over a day. Their brightness can vary by between 0.2 and 2 magnitudes.

Sagittarius

The Archer, a constellation on the ecliptic lying in the direction of the galactic center (*see* **Milky Way**). Sagittarius is the ninth sign of the **zodiac**.

satellite

An object in orbit around a planet.

satellite galaxy

A galaxy, possibly a dwarf elliptical or a dwarf irregular, that is in orbit around a much larger galaxy. For example, both the Magellanic clouds are satellite galaxies of the **Milky Way**.

Saturn

The sixth planet in the Solar System. Saturn is immediately distinct from the other planets because of its impressive system of rings. The rings are rather like a mini-asteroid belt and are composed of dust particles and pebbles. The two most obvious gaps between the rings are known as the Cassini division and the Encke division. These are the result of the action of the gravity of nearby moons. Saturn itself is a gas giant planet and a member of the outer Solar System. Its diameter is nine times greater than that of the Earth. Although the cloud features are relatively unimpressive on Saturn, spectacular outbursts are possible. These include huge storms that periodically appear on the planet every 30 years or so. The internal composition of Saturn is broadly similar to that of **Jupiter**. It does, however, have the curious distinction of existing at a mean density less than that of water. Saturn possesses 18 known moons, most of which are small rocky or icy bodies. The moon Titan however, is an interesting world because its atmosphere contains primitive organic molecules. Many astronomers believe that the atmospheric composition of Titan today mirrors that of the primeval Earth.

Schwarzschild radius

The radius within which a celestial object must be compressed if it is to become a **black hole**. Its value is determined by the mass of the object in question: the more massive the object, the less it needs to be compressed and hence the larger the Schwarzschild radius. The Schwarzschild radius gives the distance of the **event horizon** from the center of the black hole, known as the singularity.

Scorpius

The Scorpion, a medium-sized constellation on the ecliptic; the eighth sign of the zodiac. Scorpius contains the bright star Antares.

Seyfert galaxy

A specific class of galaxy that has **active galactic nuclei**. These can be **spiral galaxies** or **barred-spiral galaxies** with bright, compact galactic nuclei. The nuclei far outshine the spiral arms of the host galaxies and are far too bright to be composed only of stars. The electromagnetic radiation released by Seyfert nuclei is not in the form of **black body radiation** and cannot be starlight. These observations have led to the idea that there is something else, which is very energetic, in the nuclear region. The spectra of the nuclear regions contain spectral emission lines that suggest that hydrogen clouds are swirling around the active galactic nucleus at very high velocities. In many ways, Seyferts resemble **quasars** but are slightly less powerful and closer to us. *See also* **unified theory of active galaxies**.

CONNECTIONS

ACTIVE GALAXIES **96**
ENERGY MACHINE **98**

---

shell galaxy

An elliptical galaxy surrounded by a thin shell of stars that are thought to have been ejected during a galaxy merger. Such galaxies were discovered in 1979 by David Malin. Shell galaxies differ from **ring galaxies** in that the shells surrounding the elliptical galaxies are much farther away from the galaxy's center than the rings and much fainter. Spectroscopy of the stars in the shell show that they are old, whereas the stars in a ring galaxy are young. *See also* **ring galaxy**.

shock emission

Electromagnetic radiation, released as spectral emission lines, that is stimulated by shock waves moving through the interstellar medium. Examples include the emission coming from **Herbig–Haro objects** and some **supernova remnants**.

shock wave

A wave of compressions and rarefactions propagating through a medium at a velocity that exceeds the velocity of sound in the medium.

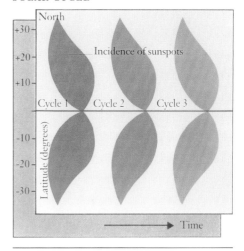

singularity

A point of infinite density which, when placed on the spacetime continuum, produces an infinitely deep potential well. Singularities are, in theory, to be found in the center of black holes. They cannot be observed, however, because they are surrounded by the black hole's **event horizon**.

solar corona

The "atmosphere" of the Sun, which extends for many times the solar radius and eventually merges with the interplanetary medium. It is very tenuous, having an average density of one-thousand billionth of that of the Earth's atmosphere. It is very hot: the gas atoms move with velocities that indicate a temperature of 2 million kelvin. The corona is visible during total eclipses of the Sun as a halo of light, mainly reflected from the **photosphere**. The corona emits ultraviolet light and X rays but is not uniform in density. It changes shape through the solar cycle.

solar cycle

The variation in the Sun's activity during an 11-year period. It is most noticeable because of its effect on the number of **sunspots** visible on the photosphere. At the start of the cycle, known as the solar minimum, there are no sunspots for several weeks. They then appear at mid-latitudes, 30° to 40° on either side of the equator. During the next five years, the sunspots continually appear and disappear, moving toward the solar equator and displaying an overall increase in number as they do so. After reaching a maximum, the sunspots continue to move toward the equator but decrease in number over the next six years. The sunspots finally arrive at approximately 7° latitude and gradually disappear. They begin to reappear at 30° to 40° again but with a reversal in their magnetic polarity. All solar activity is linked to the solar cycle.

KEYWORDS

solar nebula

The cloud of material from which the Solar System was formed. The collapse of the solar nebula into the Solar System is thought to have been triggered by the explosion of a nearby **supernova** just over 4.6 billion years ago. The material remaining after formation dissipated when the Sun passed through the **T Tauri phase**.

Solar System

Everything that is dominated by the Sun's gravitational field. The Solar System is made up of the Sun, the nine planets and their moons as well as minor bodies such as **asteroids** and **comets**. It was formed from the solar nebula 4.6 billion years ago .

solar wind

A steady stream of **plasma** flowing away from the Sun along magnetic field lines that lead into the interplanetary medium. The solar wind is composed of charged particles such as protons, electrons and atomic nuclei of helium. When these particles become trapped in the Earth's magnetic field they cause **aurorae**.

space

The volume in between stars, planets and galaxies, containing usually tenuous matter. *See* **interplanetary medium**, **interstellar medium** and **intergalactic medium**.

spacetime continuum

A geometric framework made up of three dimensions of space and one dimension of time. Within this it is possible to locate any event in terms of space and time and determine the relationships between events. The spacetime continuum is essential to the understanding of **general relativity**. In this geometrical framework gravity is explained by a curvature of the continuum.

SOLAR WIND

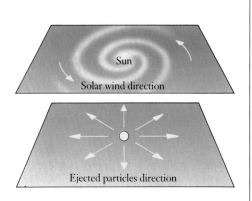

special relativity

A theory, developed by Albert Einstein and published in 1905, which describes the Universe as seen by observers in unaccelerated frames of reference. It states that everything is measured relative to something else and that the speed of light always has the same value, whatever the motion of the observer performing the measurement. The effects predicted by special relativity are **time dilation**, **mass dilation** and **length contraction**. A consequence is the equivalence of mass and energy as described by the equation $E = mc^2$. The accelerating frame of reference is dealt with by **general relativity**.

CONNECTIONS

SPECIAL RELATIVITY **60**

GENERAL RELATIVITY **72**

---

spectral absorption lines

The dark lines in a continuous spectrum that are produced when **electromagnetic radiation** passes through a gas cloud and certain wavelengths are absorbed by the electrons around the gas atoms.

spectral classification

A method of classifying stars that is based on the appearance of the spectral absorption lines in their spectra. Originally stars were classified as A to Q but better spectroscopy has led to regrouping and reorganization. The classes now run OBAFGKM. Finer divisions are provided by 10 subclasses within each spectral type denoted as 0 to 9. Spectral classification is cross-referenced by luminosity classes. There are also suffixes to identify a star's peculiar features. For example a suffix "wk" denotes that the star's spectral absorption lines are weak. Spectral type is one of the quantities plotted on the **Hertzsprung–Russell diagram**.

CONNECTIONS

COLORS AND SPECTRA **108**

BINARY AND MULTIPLE STARS **112**

EXTRATERRESTRIAL LIFE **146**

---

spectral emission lines

A very small range of wavelengths, over which the intensity of **electromagnetic radiation** is much greater than the surrounding, continuous spectrum. Emission lines are produced when electrons in gas clouds drop into lower-energy quantum states and release photons that carry away that excess energy. A spectrum with just spectral emission lines is called a line spectrum.

SPECTRUM

Continuous spectrum

Photon emitted

Emission spectrum

Photon absorbed

Absorption spectrum

---

spectral line

See **spectral absorption** and **spectral emission lines**.

spectral type

See **spectral classification**.

spectrometer

A device that splits light, such as that from a celestial object, into a spectrum.

spectroscopic binary

A **binary star** system in which the stars are too close to one another to be observed as separate stars. However, when their **spectra** are studied, the two types of **spectral emission lines** indicate that two separate stars are present. The spectral lines also changes wavelength, because of the Doppler effect, as the stars orbit one another.

spectroscopy

The study of a celestial object by investigation of its spectrum. Analysis of the spectral lines gives information about the temperature and the chemical composition of an object; and if the lines are shifted to the red end of the spectrum of a galaxy it can also indicate its distance.

spectrum

The distribution, by wavelength, of an electromagnetic radiation ray. White light is composed of the colors red, orange, yellow, green, blue, indigo and violet. The spectrum also continues into the ultraviolet and infrared. A spectrum can contain **spectral absorption lines** and **spectral emission lines**.

speed of light

The speed at which photons of electromagnetic radiation travel in a vacuum. The value is equal to 3×10^8 m/sec. According to the theory of **special relativity**, nothing can travel faster than this. *See* **universality of speed of light**.

CONNECTIONS

SPECIAL RELATIVITY 60

GENERAL RELATIVITY 72

spin

The intrinsic **angular momentum** of a subatomic particle. It is a quantum property of all subatomic particles and one of the quantum properties that defines a quantum state. **Fermions** have half-integer spins (for example, electrons have spin of one half) whereas **bosons** have integer spins.

spiral arms

Arms made up of O and B type stars that spiral outward from the galactic nucleus of **spiral** and **barred-spiral galaxies**. *See also* **density wave theory**.

spiral galaxy

Any galaxy in which a central bulge of older stars is surrounded by a flattened **galactic disk** containing a spiral pattern of young, hot stars. *See* **spiral arms** and **Hubble tuning fork diagram**.

CONNECTIONS

CLASSIFICATION OF GALAXIES 88

STRUCTURE OF GALAXIES 90

THE MILKY WAY 92

INTERACTING GALAXIES 100

star

A celestial object that shines because of the release of energy created in its core by nuclear fusion. A star can also be the hot stellar remnant left following the cessation of fusion processes in the object's core. Stars are classified according to spectral type and placed into luminosity classes. Half of them are in either **binary star** systems or multiple star systems. Some are **variable stars**, which change their size and luminosity. Stars that do not accrete enough matter to begin fusion are called brown dwarfs. Remnants of dead stars are **white dwarfs**, **neutron stars** and **black holes**. Stars assemble into collections known as **galaxies**.

starbow

The phenomenon of severe aberration of starlight that would be experienced by an observer traveling at speeds close to the speed of light. Stars appear to crowd downward in front of and behind the observer. Those in front turn blue from the **Doppler effect** and those behind turn red.

starburst

A massive bout of star formation within a galaxy. It can often be triggered by **giant molecular clouds** colliding during galaxy mergers. Electromagnetic radiation is emitted copiously at **infrared** wavelengths.

star cluster

A loose association of stars in the galactic disk of the **Milky Way**. The stars, having formed from the same dense core, are moving apart as they orbit the galactic center. Examples include the Seven Sisters (Pleiades) cluster and the Hyades cluster.

CONNECTIONS

THE MILKY WAY 92

CLUSTERS AND VOIDS 94

state of matter

A specific arrangement of atoms that is distinct from any other that a substance might possess. For example, water can exist in three obvious states of matter: as water vapor (a gas), as water (a liquid) and as ice (a solid). A fourth phase of matter is a **plasma**. The change from one state into another is known as a **phase change**.

stationary limit

The boundary of the **ergosphere** around a black hole.

stellar core

The central region of a star in which nuclear fusion takes place. When a star dies, the core is often all that is left. The nuclear fusion will then have stopped.

stellar population

Two broad categories of star populations. Population II stars are found in **globular clusters**, in **elliptical galaxies** and also within the galactic nuclei of **spiral galaxies**. They are all old stars with a low metal content. Population I stars are young stars with a high metal content and are found in the spiral arms of spiral galaxies.

CONNECTIONS

GALAXIES AND QUASARS 86

VARIABLE STARS 114

stellar wind

A steady stream of plasma flowing away from a distant star along magnetic field lines that lead into the interstellar medium. It is composed of charged particles such as protons, electrons and the atomic nuclei of helium.

strong nuclear force

The force of nature through which hadrons communicate with each other. It is the strongest of all the forces but it acts only over the distance of an atomic nucleus. In the **grand unified theory** (GUT), the strong nuclear force is linked to the **weak nuclear force** and to **electromagnetism**.

subatomic particle

Any particle that is smaller than an atom. The term is sometimes used to refer to the particles protons, neutrons and electrons, which make up atoms. Subatomic particles are also found in **cosmic rays**. Many particles require very high energies to exist, and decay in fractions of a second.

Sun

The star at the center of the Solar System, around which the Earth and other planets orbit. It is a **yellow dwarf** and has a spectral classification of G2. It is about 4.6 billion years old and contains 2×10^{30} kg of material. Its diameter is over 100 times that of the Earth, having formed by accretion in the gravitational collapse of the **solar nebula**. Products of radioactive decay in **asteroids** indicate that the gravitational collapse was possibly triggered by a nearby supernova. *See also* **chromosphere**, **photosphere**, **solar cycle**, **solar wind** and **sunspots**.

sunspot

A temporarily cooler region on the **photosphere** of the Sun – typically between 1000 and 2000 degrees below that of the photosphere, which is approximately 6000 K. Sunspots are thought to be caused by a magnetic field bursting through the photosphere. They often appear in pairs or groups. In a pair, one spot has north magnetic polarity and the other south. They become more prevalent every 11 years, and migrate though latitudes, moving closer to the equator as the **solar cycle** progresses.

supercluster

A cluster of **clusters** of galaxies that stretches for hundreds of millions of light-years. So far about 50 such superclusters have been identified.

superforce

A hypothetical force that is thought to incorporate all the forces of nature: the **strong nuclear force**, the **weak nuclear force**, **electromagnetism** and **gravity**. It is predicted by unified field theories and could have been present in the Universe only during the Planck time – that is, the first 10^{-43} seconds after the **Big Bang**.

supergiant

A star which has a higher luminosity and a larger radius than a giant of the same **spectral classification**. A typical supergiant has up to 100 times the luminosity of a giant star and almost always results in one form of **supernova**.

supernova

A catastrophic explosion that blows a star to pieces. There are two types. Supernovae type I are thought to be **white dwarfs** in binary star systems which accrete material in a similar way to novae but, instead of there

SUPERCLUSTER

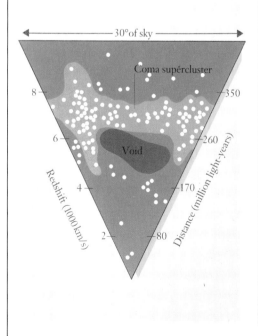

being a **nova** outburst, the material builds up until the resulting nuclear fusion is so powerful that it disrupts the entire star. Type II supernovae are stars greater than eight solar masses. The nuclear fusion in the core of these stars produces all the elements in the Periodic Table up to and including iron. Iron fusion is not a reaction that gives out energy, so the fusion process can go no further and the iron core is therefore inert. When it reaches 1.4 solar masses (the **Chandrasekhar limit**) gravity overwhelms the electron degeneracy pressure, the stellar core collapses and the star collapses down onto it. The impact of the material onto the collapsed core produces a shock wave which, in turn, disrupts the star and blows it apart. Type I supernovae are even more luminous than type II but both are very luminous and each outshines the galaxy in which it is contained. A supernova fades over a period of months, as the original energy of the explosion is replaced by the radioactive decay of unstable atomic nuclei. If it is not destroyed altogether, the central core is left as a **neutron star** or as a **black hole**. The outer material is blown off into space to form a nebula that is often referred to as a supernova remnant.

supernova remnant

A nebula of glowing gas created either by ejecta from a **supernova** radiating excess energy into space, or by the shock emission caused by shock waves, produced in the explosion, traveling through the interstellar medium.

supersymmetry

A set of rules used in some **unified field theories** that state that all fermions have boson counterparts and vice versa, so the two are linked. For example, a photon has a fermion counterpart, the photino; and an electron has a boson counterpart, the selectron.

synchrotron

A particle accelerator for bringing pulses of charged particles to very high energies through a circular field, before colliding them with a target to observe the result, using a variety of detectors.

synchrotron emission

The emission of **electromagnetic radiation** by electrons spiraling around the lines of magnetic fields.

T Tauri phase

A phase through which low-mass, **pre-main sequence** stars pass. The onset of nuclear fusion in the core of a young star brings about unstable pulsations and strong **stellar winds**. Unstable stars at this stage of their evolution are known as T Tauri stars. This phase is often associated with **Herbig–Haro objects** because the powerful outflows collide with the surrounding interstellar medium and cause shock emission. Collections of T Tauri stars are known as T Tauri associations.

tauon

A subatomic particle, a cousin of the electron, like the muon, but heavier. A tauon is a negatively charged lepton. It is 3560 times more massive than an electron.

Taurus

The Bull, a large constellation on the ecliptic; the second sign of the **zodiac**. It contains the Crab nebula, the star clusters the Hyades and Pleiades, and the bright star Aldebaran.

telescope

An instrument that can collect, focus and magnify starlight. Different designs of telescope use mirrors or lenses to bring the light to a focus. Some specialized designs employ a combination of both. Reflecting telescopes use mirrors, refracting telescopes use lenses and catadioptic telescopes use both. Telescopes are mounted on either **altazimuth mountings** or **equatorial mountings**.

temperature

A property that measures the effect of heat in an object. This can be measured in Celsius or kelvin degrees. Fahrenheit is also used for everyday purposes.

terrestrial planet

Any one of four rocky planets found within the inner Solar System. They are Mercury, Venus, Earth and Mars. The terrestrial planets are distinguished from the gas giants (the outer planets: Jupiter, Saturn, Uranus, Neptune and Pluto) by their smaller diameters, higher densities and their tenuous atmospheres.

thermodynamics

A set of rules contained within the laws of physics stating how heat flows between objects at different temperatures. The first law

TIDE

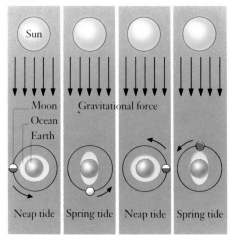

Sun

Moon
Ocean
Earth

Gravitational force

Neap tide   Spring tide   Neap tide   Spring tide

of thermodynamics states that energy can neither be created nor destroyed, heat and mechanical energy being mutually convertible; the second law, that it is impossible to convey heat from a cooler body to a warmer one without work being done; and the third law, that it is impossible to devise any set of operations that will achieve a temperature of **absolute zero**.

tidal tail

A tail-like structure of stars flung out behind galaxies when they undergo a galaxy merger. The tails are produced when the galaxies fall into each other's **potential wells**.

tide

The deformation of the spherical shape of a celestial object (including the Earth's covering of seawater) brought about by the gravitational attraction of another object nearby.

time

A fundamental property of the Universe. It indicates the direction in which energy will flow during a chemical reaction. For example, in **thermodynamics,** heat will always flow from a hotter object to a cooler one. If it were to flow in the reverse direction, time would have to be traveling backward, which is never observed to happen. Relativity predicts that time passes at different rates depending on the strength of the local gravitational field and the speed at which the observer is traveling. Despite these strange effects, time always flows forward.

time dilation

The relative slowing down of time within a frame of reference approaching the speed of light, compared with the surrounding Universe. This effect is predicted by special relativity. Time will also pass more slowly in a strong gravitational field than in a weak one. *See also* **length contraction** and **mass dilation**.

triple alpha process

A nuclear fusion process that converts helium to carbon. Two atomic nuclei of helium (alpha particles) collide and form beryllium. During a third collision carbon is formed. Successive collisions build oxygen and neon. This process powers **red giant** and red supergiant stars.

ultraviolet

A type of **electromagnetic radiation** with a wavelength shorter than that of visible light. It is defined to be within the range of 10 to 400 nm. Electromagnetic radiation with a wavelength shorter than 10 nm includes X rays and gamma rays.

unified field theory

Any theory that explains the **strong nuclear force**, the **weak nuclear force**, **electromagnetism** and **gravity** as different aspects of the same superforce. In theory, it is supposed to have been present in the Universe only during the Planck time, when the Universe was less than 10^{-43} seconds old. Following this, gravity immediately became independent of the other three forces. They continued to function as the **grand unified force** until later in the history of the Universe, when they too separated, in stages, into the forces we know today.

unified theory of active galaxies

Any theory that seeks to explain how two or more types of active galaxy are actually the same class of object observed at different orientations to the Earth. An example is a scheme which aims to link two types of Seyfert galaxies. The difference between these types of galaxy is in the width of the spectral emission lines. These differentiate the speed at which the hydrogen clouds are orbiting the active galactic nucleus. According to the unification theory, it is just a matter of the Seyfert galaxy's orientation to us, which determines whether observers on the Earth can see into the deep regions of the nucleus where the faster-moving clouds exist. If this region is blocked from our view, then observers see only the slower clouds. Most unification schemes assume that the active galactic nucleus is surrounded by a torus of dusty material, which blocks our

view when the galaxy is oriented edge-on to us. Grand unification schemes attempt to link together all active galaxies, such as **Seyfert galaxies**, **quasars** and **BL Lacertae** objects. There is currently great debate among astronomers over whether radio-loud active galaxies, such as quasars and radio galaxies, can be unified with radio-quiet active galaxies such as Seyferts.

universality of the speed of light

The principle that the speed of light is the same whether the observer measuring it is stationary or in constant motion. It is one of the cornerstones of **special relativity**. It was first proved experimentally by the US physicists A.A. Michelson and E.W. Morley in 1887, when trying to detect the ether, the hypothetical medium, believed to pervade the whole of space, through which electromagnetic radiation supposedly propagated. In an attempt to measure the difference in light's velocity through this ether, it was shown that all experiments return the same value for the speed of light, regardless of the relative velocity of the experiment. At a stroke, this disproved that the ether existed and paved the way for Albert Einstein to develop the theory of **special relativity**.

Universe

The whole of space, time and everything in it. It was created during the Big Bang and is expanding. It is assumed by astrophysicists that the laws of physics apply equally in all parts of the Universe. It is possible that other universes exist or that, if our universe is an oscillating one, there may have been prior universes and may be subsequent ones, in which different physical laws could apply.

Uranus

The seventh planet in the Solar System and one of the gas giant planets of the outer Solar System. It has a similar composition to that of Neptune, namely a significant proportion of icy material as well as the plethora of gases expected from a gas giant planet. When compared with the other gas giant planets, Uranus' atmosphere and cloud structure are remarkably quiescent. There are virtually no visible cloud belts and certainly no storms or cyclonic systems. It is thought that this is because Uranus, unlike the other gas giant planets, does not have a significant source of internal heat. Uranus is notable for the fact that its rotation axis lies almost in the plane of the ecliptic. This is unlike the other planets whose rotation axes are roughly perpendicular to the plane of the Solar System. This curious alignment means that Uranus appears to roll through sections of its orbit. Uranus has 15 known moons, 10 of which are small rocky or icy objects. Of the five larger objects, Miranda may have once been almost shattered by a collision with another celestial object.

variable star

Any star whose luminosity varies. The variations can be intrinsic, because of internal processes, or extrinsic, due to eclipses and other phenomena. They can also be irregular or periodic. If periodic, the timescale may be just a few hours or it may be several years. There are many kinds and classifications of variable star. For example, Mira-type variables have relatively long periods but Cepheid variable stars have periods of between one day and one hundred days. Irregular variables include **novae** and **supernovae**, which are sometimes referred to as cataclysmic variables. Other irregular variable stars are those currently in their **T Tauri** phase. Extrinsic variables are caused by the rotation of stars with non-uniform brightness or by eclipses in **eclipsing binaries**. The variation in brightness of all forms of variable star can be plotted on a graph of brightness against time. Such a graph is known as a light curve.

Venus

The second planet in the Solar System in order of distance from the Sun. Venus is one of the terrestrial planets and forms part of the inner Solar System. It is slightly smaller than the Earth, with an atmosphere so dense that it is permanently covered in clouds. For this reason, the surface is always shielded from direct view, and the clouds can be penetrated only by radio waves. The atmosphere is composed almost entirely of carbon dioxide, which causes the planet to retain the heat that it receives from the Sun. This intense greenhouse effect means that the surface of the planet is at the incredibly high temperature of 450°C. In the higher layers of the planet's atmosphere, carbon dioxide is changed by the Sun's electromagnetic radiation. It reacts with the small amounts of sulfur and water present to form droplets of dilute sulfuric acid. These droplets then condense and fall as rain onto the planet. The surface of Venus shows definite evidence of past volcanic activity but whether this is still taking place is uncertain. Venus has no natural satellites.

Virgo

The Virgin, a large constellation on the ecliptic; the sixth sign of the zodiac. It contains the bright spectroscopic binary (*see* **double stars**) Spica and a cluster of galaxies.

virtual particles

Particles of matter (together with their anti-matter counterparts) that spontaneously spring into existence and then annihilate one another in accordance with **Heisenberg's uncertainty principle** and the equivalence of mass and energy ($E = mc^2$). The more massive the particles, the shorter the length of time they can exist. Virtual particles carry the fundamental forces between "real" elementary particles. *See* **Hawking radiation**.

CONNECTIONS

EXCLUSION AND UNCERTAINTY 66

FORCES AND FIELDS 68

LONG-TERM FUTURE 136

---

visible light

The range of **electromagnetic radiation** to which our eyes are sensitive. We perceive it in colors, ranging from violet, at approximately 400 nm in wavelength, through yellow, at 500 nm, to red, at 700 nm. The range is bounded by ultraviolet, which has a wavelength of less than 400 nm and infrared, whose wavelength exceeds 760 nm.

visible spectrum

A spectrum consisting only of visible light.

visual binary

A double star that appears to be **binary star** system because both stars happen to lie in the same direction. They are not, in fact, physically associated and so are not true binary stars. They simply appear close together on the **celestial sphere** because of a chance alignment.

VLBA (Very Long Baseline Array)

An **interferometer** stretching all the way across North America. Its most extreme components are in Hawaii and in St. Croix in northeast Canada. This provides a radio telescope with an effective diameter of 8000 km and gives a resolution of 0.2 millionths of an arcsecond.

VLT (Very Large Telescope)

A project by the European Southern Observatory to construct and link four 8-meter telescopes at a site in South America. Once the project is completed, the combined telescopes will be the most powerful in the world.

void

A vast and roughly spherical region of space that is surrounded by **superclusters** of galaxies. The regions themselves contain hardly any galaxies, though one or two have been found. Voids are connected to each other in a similar way to the holes in a natural sponge. Why the Universe should have this sponge-like structure is one of the greatest puzzles facing modern astronomers.

CONNECTIONS

SCALE OF THE UNIVERSE 76

INFANT UNIVERSE 82

CLUSTERS AND VOIDS 94

W particle

A subatomic particle, discovered in 1983 at CERN in Europe, thought to be, with the Z particle, the carrier of the weak nuclear

WAVELENGTH

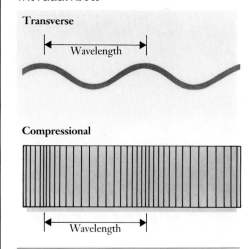

VARIABLE STAR

force. It was found in both positive and negative forms. Its discovery provided crucial evidence in favor of the **electroweak force**.

W Virginis star

A **Cepheid variable** star of stellar Population II. These stars are two magnitudes fainter than a classical Population I cepheid.

waterhole

A region of the electromagnetic spectrum, consisting of microwaves of between 1000 and 10,000 MHz, at which radiation from space and noise from the Earth's atmosphere are least. The region is heavily scrutinized by researchers seeking radio transmissions from other stars. The region is bordered by spectral lines for the ions H^+ and OH^-

wave function

A mathematical quantity that describes particles as a wave in **wave-particle duality**. Its magnitude is analogous to the amplitude of electromagnetic radiation: a measurement of the disturbance caused by the ray's propagation through an electromagnetic field. The wave function defines the probability that the particle is located in that specific region of the **de Broglie wavelength**.

wavelength

The measurement of the distance from crest to crest, or trough to trough, in transverse waves such as **electromagnetic radiation**. It is linked to frequency. In waves propagating with the same velocity, the longer the wavelength, the less the frequency.

wave-particle duality

A fundamental concept of quantum theory whereby waves can behave like particles and particles can behave like waves. For example, electromagnetic radiation can be thought of as either a wave or a particle (a **photon**) but never as both at the same time. Considering particles such as electrons as waves can lead to an understanding of why they exist in only specific orbits around atomic nuclei.

weak nuclear force

The force of nature through which neutrinos interact. It is much weaker than the **strong nuclear force** and underlies **beta decay** radioactivity, in which a neutron emits an electron and becomes a proton. It is linked to the electromagnetic force at high energies to create the **electroweak force**.

white dwarf

A highly evolved stellar remnant, the remains of a stellar core after nuclear fusion ceases. It is composed of **electron degenerate matter** and is often found in planetary nebulas. Its temperature can be between 100,000 and 4000 kelvin. White dwarfs do not have an internal heat source but they gradually radiate their residual energy until they become **black dwarfs**. They are very small, sometimes with a diameter less than that of the Earth. Their upper mass limit is 1.4 solar masses (the **Chandrasekhar limit**). If it were any larger than this, the white dwarf would become a **neutron star**, composed of baryon degenerate matter. White dwarfs in **binary star** systems can produce **novas** and type I **supernovas**.

CONNECTIONS

KINDS OF STARS 102

GIANTS AND DWARFS 110

POST-MAIN SEQUENCE 124

Wien's law

A law of physics describing the wavelength of peak emission of **black body radiation** as a function of temperature. The higher the temperature, the shorter the wavelength at which maximum spectral radiancy occurs.

Wolf–Rayet star

An exceptionally hot star with a surface temperature between 20,000 and 50,000 K. It is characterized by **emission lines** in its spectrum. This is because it is surrounded by large clouds of gas, produced as it undergoes periods of intense mass loss in the form of **solar winds**. Sometimes such stars are found at the center of planetary nebulas that are still forming.

worm hole

A hypothetical shortcut through the spacetime continuum. Worm holes, if they exist, could be used to permit interstellar travel to apparently remote regions of the Universe.

WIEN'S LAW

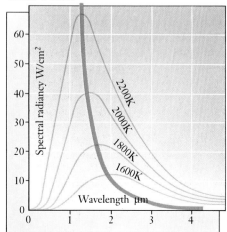

X-ray astronomy

The study of the X rays emitted by celestial objects. Because the Earth's atmosphere absorbs most X rays, X-ray astronomy is usually done from high-altitude balloons and satellites. X rays are normally emitted by regions of superhot gases. Sources include the solar corona, solar flares, compact stars in binary systems, quasars and intergalactic gas.

X rays

High-energy **electromagnetic radiation** of wavelengths in the range 10 to 0.1 nm. The only photons that are more energetic than these are the **gamma rays**.

yellow dwarf

Any star with a spectral classification of G on the **main sequence**; for example, the Sun.

Z particle

A subatomic particle, discovered in 1983 at CERN in Europe, thought to be, with the W particle, the carrier of the weak nuclear force. Its discovery provided crucial evidence in favor of the **electroweak force**.

zenith

The point on the **celestial sphere** directly above the observer and at right angles to the celestial horizon, directly opposite the nadir.

zero-age main sequence

The position on the main sequence of a **Hertzsprung–Russell diagram** reached by stars after they have passed through their **T Tauri phase**. It corresponds to a star with 75 percent hydrogen, 23 percent helium and 2 percent metals in its core. Nuclear fusion gradually changes these proportions and the star evolves across the main sequence.

zodiac

The band of the heavens whose outer limits lie 9° on each side of the ecliptic. The 12 main constellations near the ecliptic, corresponding to the signs of the zodiac, are **Aries, Taurus, Gemini, Cancer, Leo, Virgo, Libra, Scorpius, Sagittarius, Capricornus, Aquarius, Pisces**. The orbits of all the planets except Pluto lie within the zodiac and their positions, as that of the Sun, are important in astrology. The signs are each equivalent to 30° of arc along the zodiac.

zodiacal dust

Dust found in the inner Solar System, in the plane of the Solar System, which is believed to have been created by asteroid collisions and the breaking up of comets. The dust causes light to be reflected and this sometimes shows up as a cone of light along the ecliptic during the hours of twilight.

A UNIVERSE
of Rules

FROM ITS SMALLEST PARTICLE to its largest galaxy, everything in the Universe follows rules that are described by the laws of physics. Formulating these laws has traditionally been the job of physicists, whereas charting and cataloguing the heavens has been done by astronomers.

With the realization that laws of physics should apply to the entire universe, a new type of scientist has emerged: the astrophysicist. An astrophysicist uses the observations of astronomers and the rules of physicists. In this way the phenomena and objects in the Universe can be explained in terms of physical laws that have been tested in Earth-bound laboratories.

All matter in the Universe is made up of five stable "elementary particles": electrons, protons, neutrons, neutrinos and photons. There may also be a sixth elementary particle known as the graviton. The first three particles make up the entire content of the visible material in the Universe; the other two, and the hypothetical graviton, carry energy created by interactions between the first three. These interactions are caused by the forces of nature. There are four fundamental forces: the strong nuclear force, the weak nuclear force, electromagnetism and gravity. The interactions between every object in the Universe can be explained in terms of these four forces.

The extinct volcano Mauna Kea, in Hawaii, is one of the best sites from which to observe the Universe. Astronomers from many countries have built telescopes here which allow them to probe the heavens in unprecedented detail. In this time-lapse photograph, the total solar eclipse of July 11, 1991 has been caught on film. The very fact that we can predict such events is a testimony to the understanding we have gained from studying the heavens and the wonders they contain. This understanding is based in turn on the precise observation and measurement demanded by modern science.

INSIDE THE ATOM

EVERY object in the Universe, even the most distant star, is made of atoms. Although the word originally meant "indivisible", and atoms were thought to be the smallest units of matter, physicists in the 20th century have discovered that atoms themselves are made up of separate components.

Atoms are complex structures which have a small central region, the nucleus, where the protons and neutrons reside. Electrons occupy orbits around the nucleus. The atom itself is held together by two of the fundamental forces of nature. The strong nuclear force causes the protons and neutrons to be powerfully attracted and to hold together to form the nucleus. The strong nuclear force is the strongest force of nature, but it is very limited in scope: it can be felt only on the scale of an atomic nucleus, about 10^{-14} m. Gravity, a more familiar fundamental force, is the opposite: by comparison with the strong nuclear force it is very weak, but it is felt across huge distances throughout the Universe.

Protons carry a positive electric charge, whereas the neutrons, as the name implies, are electrically neutral. As a result, regardless of the number of protons or neutrons, the nucleus has an overall positive charge.

The electromagnetic force holds the electrons in orbit around the nucleus. Electrons are negatively charged, and they are impelled by the electromagnetic force to orbit the positively charged nucleus. Because there are normally as many electrons as protons in an atom, they cancel each other's electrical charge, so that the atom's overall atomic charge is usually zero.

The number of protons in the nucleus determines which chemical element the atom will be. For instance, a nucleus containing a single proton is hydrogen. Different numbers of protons give rise to other chemical elements, such as helium (2 protons), iron (26 protons) and uranium (92 protons). These elements have very different properties from each other. The sum of the protons and neutrons gives the atomic mass. Hydrogen is the lightest element, and uranium the heaviest naturally-occurring element. Most of the mass of an atom comes from its protons and neutrons; electrons have negligible mass.

▽ In 1919, Niels Bohr proposed a "model" of the atom. Electrons in circular orbits surrounded a central nucleus. Bohr envisioned that the electrons were like negatively-charged "balls". The nucleus, in the center, is a complex environment with its own structure. Nuclei contain positively-charged protons and neutral neutrons (which do contribute mass). Hydrogen's nucleus contains only a single proton.

The number of protons in the atoms of a particular element is fixed, and is unique to that element. The number of neutrons may vary. This is especially true in larger, heavier atoms, in which the number of neutrons may be greater than the number of protons. In heavy atoms, the protons are crowded together and tend to repulse each other. Extra neutrons neutralize this effect. Without them, the nucleus of a very heavy atom would blow itself apart. Uranium, for instance, has 146 neutrons and 92 protons. Anything with more protons must be made in a laboratory.

Uranium also occurs in natural form with 143 neutrons. Such different combinations of an element are called its isotopes. Some isotopes – such as uranium-235 and 238 – are unstable and undergo radioactive decay. A radioactive isotope emits particles or energy in order to restore its stability. It may lose protons, or one of its neutrons may turn into a proton. Either way, the number of protons in the nucleus changes, and the element usually becomes a different element in the process.

The nuclei of some atoms split in two in the process known as nuclear fission. This occurs predominantly with elements that contain more than protons than iron (which has 26) – known as heavy elements. Fusion – a joining of nuclei – can take place between isotopes of elements lighter than iron. Fusion is the reaction that powers stars.

In both fission and fusion, the result is a different chemical element which may or may not be stable. Fusion and fission can release vast amounts of energy because the mass of the resulting component particles is slightly less than the sum of the masses of the original particles. It is this difference in mass, known as the mass defect, that is converted into energy.

△ **Located on the plains of San Augustine, New Mexico, the Very Large Array telescope is a collection of 27 identical radio telescopes which work together in order to detect high-frequency radio emission from atoms and subatomic particles in space. Neutral hydrogen** atoms can emit radio waves with repeating patterns every 21 centimeters. This emission can be picked up by these telescopes and provide enough data to draw maps of the density of hydrogen in the Galaxy. Other forms of radio waves, from other kinds of atoms, can also be detected.

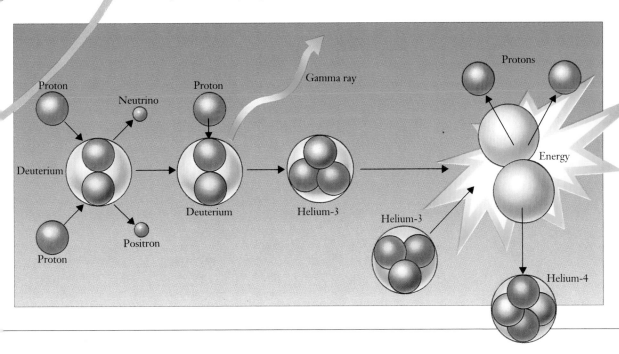

◁ **Two hydrogen nuclei fuse together under conditions of extreme temperature, extreme pressure, or both. The process of hydrogen fusion eventually leads to the formation of helium. This reaction is known as the proton-proton chain because nuclei of hydrogen consist of single protons. It takes six hydrogen nuclei to make one of helium. In the final reaction, however, two hydrogens are returned unchanged. These may then take part in another proton-proton chain interaction.**

Proton
Neutrino
Deuterium
Positron
Proton
Proton
Gamma ray
Deuterium
Helium-3
Protons
Energy
Helium-3
Helium-4

PARTICLE EXPERIMENTS

INDIVIDUAL atoms can only be seen with the most sophisticated microscopes, and the particles that comprise them are impossible to detect by conventional means, because they are so small. However, because atomic nuclei and electrons are electrically charged, they can be controlled by magnetic fields. Using this knowledge, physicists in the 20th century have devised experiments to deduce the existence of these particles and their physical properties. The results were based upon observing the outcome of collisions between particles and atoms. The higher the speed of the colliding particles, the higher the energy of the collision and the greater the detail in which these interactions can be observed.

Understanding the subatomic structure of matter is a triumph of 20th-century physics. In 1911, the New Zealand-born British physicist Ernest Rutherford made a breakthrough while studying the results of an experiment in which positively-charged, radioactive alpha particles were fired at a thin sheet of gold. Most passed straight through the gold sheet unhindered, but some were deflected by a small amount, and a tiny proportion rebounded altogether.

Rutherford calculated that the positive charges on atoms are concentrated at the centers of their volumes, with the rest of the atom consisting largely of empty space. This "nuclear" model, therefore, established that atoms have their positive charge, as well as most of their mass, concentrated in a small region called the nucleus at their centers.

In 1913, the British physicist Henry Moseley (1887–1915) built on the work of his contemporary, the Danish physicist Niels Bohr (1885–1962), in conjunction with experimental work to understand the distribution of electrons around a hydrogen

▷ The detection of particle collisions takes place in a large, sophisticated apparatus, custom-built for the purpose. RIGHT A technician checks the end caps in the detector called **OPAL** at CERN in Geneva, Switzerland. The image FAR RIGHT shows a typical detector trace, this time from the ALEPH detector at CERN. The dotted lines delineate the paths of particles produced in the collision. A magnetic field causes them to curve which makes their identification much easier.

▽ Atomic colliders cause subatomic particles to be accelerated to high energies, and hence velocities, at which they are made to collide with other accelerated particles. Usually the collision takes place between matter, electrons perhaps, and their antimatter counterparts, positrons. In the linear accelerator to the right, electrons are produced and then accelerated along the fly-tube. The acceleration is produced by magnets which affect the negatively charged electrons. Since this would tend to produce circular motion, the magnets must be arranged so that the flight paths are straightened out. The electrons can be accelerated to close to the speed of light and then collided with the positrons in a detecting chamber. This diagram shows electron production and finishes with the detector. The positrons are produced in a duplicated system, except for reversed magnetic polarities, running on the other side of the detector.

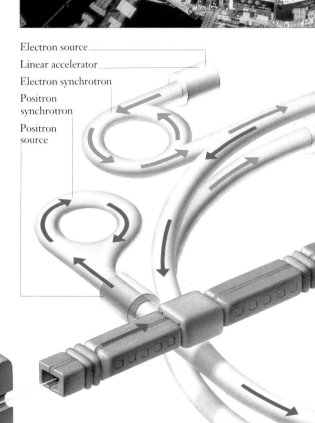

Electron source
Linear accelerator
Electron synchrotron
Positron synchrotron
Positron source

Positron source

Linear accelerator

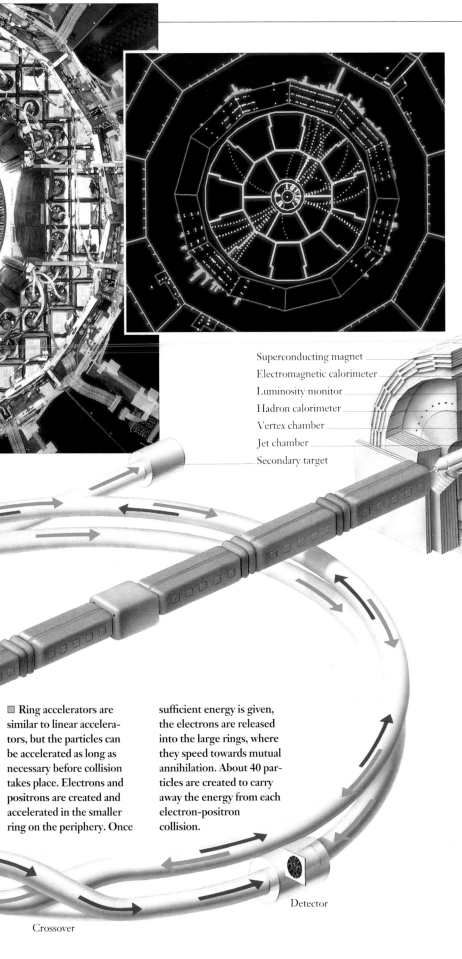

▽ Detectors must provide an area for the collision of subatomic particles to take place. They must also be able to record the tracks of ejected particles following the interaction.

The particle detector used for positron-electron collisions at CERN is known as ALEPH: an acronym for *A*pparatus for *L*arge *E*lectron *P*ositron collider p*H*ysics.

Muon detectors

Superconducting magnet
Electromagnetic calorimeter
Luminosity monitor
Hadron calorimeter
Vertex chamber
Jet chamber
Secondary target

▢ Ring accelerators are similar to linear accelerators, but the particles can be accelerated as long as necessary before collision takes place. Electrons and positrons are created and accelerated in the smaller ring on the periphery. Once sufficient energy is given, the electrons are released into the large rings, where they speed towards mutual annihilation. About 40 particles are created to carry away the energy from each electron-positron collision.

Crossover

Detector

atom. Bohr had shown that electrons orbit hydrogen nuclei in circular orbits defined by their energy. By absorbing or emitting energy, they could "jump" between orbits. Moseley further investigated this idea by studying the emission of X rays (very high-energy photons) from substances bombarded by high-energy electrons. Moseley deduced that electrons around any atom exist in energy-defined shells and can jump between them.

In recent decades, particle physicists have concentrated on investigating the interior of the nucleus, particularly the structure of the neutrons and protons. Huge machines, known as particle accelerators, are used to accelerate charged particles to velocities close to that of light passing through rings of magnets several kilometers in diameter. The particles are then smashed into others, and the results of the collisions are detected by devices such as a bubble chamber in which the tracks of fast-moving particles can be photographed. The tracks reveal the way the particles have moved through a magnetic field. This allows scientists to determine their mass and their charge, and so conclude what the particles are. Using such techniques, specialists have discovered new particles that confirm current theories about the nature of the fundamental forces in the Universe.

ELECTROMAGNETIC RADIATION

W E CAN see because our eyes are sensitive to light. Light is an electromagnetic disturbance that has many of the same properties as waves. For instance, it has crests and troughs; the distance between each crest is the wavelength. Related to wavelength and the speed at which the wave moves is the frequency. Frequency measures the number of wave crests that pass a specific point in a given time. Because light always travels through space with the same velocity, light of longer wavelengths has a lower frequency than light of shorter wavelengths.

Light of different wavelengths is distinguished by its color. Red light, for example, has a longer wavelength and lower frequency than blue light. The brightness at which we view a light source depends upon the number of photons which strike our eyes. A large number of photons means that the light source appears bright, but fewer photons make the light dimmer.

Visible light occupies only a small part of the electromagnetic spectrum. Radiation with a longer wavelength than red light – infrared radiation – is sensed by humans as heat. At even longer wavelengths are microwaves and then radio waves. On the other side of the visible spectrum, at ever shorter wavelengths, are ultraviolet, X rays and gamma rays.

KEYWORDS

ELECTROMAGNETIC SPECTRUM
GAMMA RAYS
INFRARED
IONIZATION
MICROWAVES
PHOTON
RADIO WAVES
REST MASS
SPEED OF LIGHT
ULTRAVIOLET
VISIBLE LIGHT
WAVELENGTH
X RAYS

▽ The electromagnetic spectrum (shown in the diagram below) is formed by electromagnetic radiation, naturally arranged according to its wavelength. The visible portion of the spectrum is only a very small part. Wavelength increases from the left to the right of the diagram. This means that the frequency decreases from left to right. These are the classical properties of the electromagnetic waves. According to quantum theory, each piece of electromagnetic radiation is carried by a photon. The energy of the photons is greatest at the gamma ray end of the spectrum and weakest at the radio end of the spectrum. The relative energy of the photons is shown below the spectrum by imagining that a visible light photon can travel one meter through a block of material before being absorbed. Gamma rays can travel four million times as far; radio photons only 0.0000005 times as far.

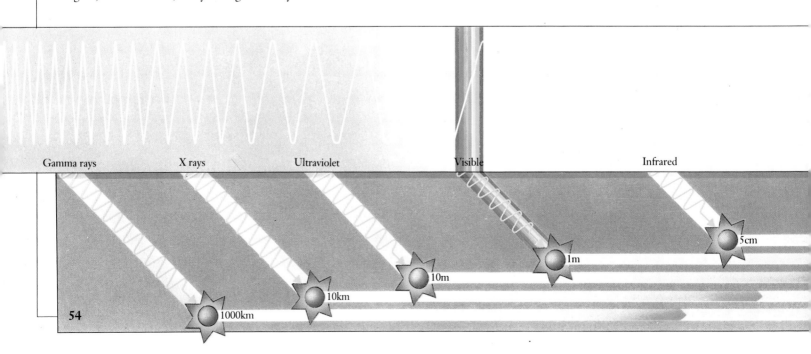

Gamma rays · X rays · Ultraviolet · Visible · Infrared

1000km · 10km · 10m · 1m · 5cm

Emission

◁ Spectral lines are produced by electrons moving between energy levels around atomic nuclei. In atomic absorption, some passing photons from an illuminating source are absorbed by the electrons. The absorbed energy corresponds to the exact energy needed by the electron to reach another level. This removes certain energies from the light passing by the atom. If it is then passed through a prism, the resulting spectrum has dark lines corresponding to the absorbed energies. The electrons at higher energy levels soon jump back down, re-emitting the photons. These photons may be emitted in any direction rather than placed back on their original paths. Observation from any line of sight which does not correspond to the original line of illumination will result in the observer seeing only a dark background with bright lines. These lines correspond to the energies that are emitted.

Fundamentally, these radiations are the same; they are distinguished only by their different wavelengths.

Shorter-wavelength radiation carries more energy than radiation of longer wavelengths. The energy of the radiation is carried by fundamental particles known as photons. They travel at the speed of light and have no rest mass. This means that if a photon which is at rest (stationary) could be weighed, it would have no mass. When the photon is moving, however, the energy of movement causes it to act as if it had mass. The energy carried by a photon is defined by the wavelength of the light which the photon represents. For example, blue-light photons carry more energy than red-light photons. Gamma rays are the photons with the highest energy, and radio waves are photons with the lowest energy.

Photons are emitted and absorbed from electrons and affect matter by interacting with other electrons. An electron needs energy to jump from its natural orbit around its nucleus to a higher one. When a photon of the right energy collides with the electron, the electron absorbs the photon and jumps to a more energetic level around the atom. However, the electron does not normally stay there for long, and soon jumps back into its original state (not necessarily in one step). When an electron does this, a new photon is emitted, which carries the energy away.

Sometimes an electron absorbs a photon that is carrying so much energy that it allows the electron to escape the atom altogether. This is known as ionization. The atom becomes positively charged, because it now has fewer negatively-charged electrons than positively-charged protons. The atom in this state is known as an ion. The escaped electron is free to wander around the Universe until it is captured by another ion that needs a spare electron.

Microwave

>0.05 mm

Radio

0.00005 mm

OBSERVING THE HEAVENS

EYES are the most familiar instrument used by people to observe the electromagnetic spectrum. In the same way as human eyes focus light, detect it and send the image to the brain for processing, so modern telescopes focus light from an image and prepare it for processing in a computer. Some optical telescopes, like the eye, use lenses to collect and focus the light from the night sky; but most large astronomical telescopes use mirrors to bring the light to a focus. Big telescopes are used because they gather much more light than smaller ones and allow astronomers to see much fainter objects. The world's largest telescope, the Keck telescope at Mount Palomar in California, has an aperture of 10 meters.

To use these telescopes to their full advantage, they are placed in high and remote places. They have to be well away from city lights with as little atmosphere to peer through as possible. This is to lessen the blurring of the light from celestial objects. A telescope in space escapes the Earth's atmosphere altogether and provides the clearest images.

Telescopes that use lenses are called refractors; telescopes that use mirrors are called reflectors. Mirrors are used for the very largest telescopes because the light can be "folded" inside the telescope tube. This means that they do not have to be as long as refracting telescopes. Newtonian designs use a curved primary mirror and a flat secondary mirror to bring the light to a focus; Cassegrain telescopes improve on this by having two curved mirrors.

Although a well-made telescope gives a perfect image if pointed directly at a celestial object, most cannot successfully focus the light from objects which are not directly in line, such as two different objects in the same field of view. This is known as off-axis aberration. Other types of aberration include chromatic aberration and spherical aberration. Chromatic aberration results from the inability of a lens to bring light of different wavelengths (colors) to the same focus. It is corrected by compound lenses using two or more different types of glass. Spherical aberration affects mirrors as well as lenses, resulting from the difficulty of grinding their curved shapes perfectly. Reflecting telescopes without aberrations can be constructed using a combination of lens and mirrors. They are known as catadioptic, and designed in such a way that the aberrations of the lens counteract the aberrations of the mirror.

From Earth, the stars appear to move across the sky. This "apparent motion" is caused by the rotation of the Earth. If a telescope is left pointing at the sky while a photograph is taken, the image is blurred, because of the passage of the stars across the night sky. Thus the mounting must also move to counteract this rotation and keep the telescope pointing directly at the object of interest. The simplest are altitude and azimuth mountings, but computers are required to track objects because both axes must be moved at once. Better designs, although more difficult to make, are known as equatorial mountings. They have their axes aligned to the equator and rotational axis of the Earth. They need only a motor to drive them around the polar axis for them to track an object.

Some of the common instruments used in astronomy are the photometer, which measures the brightness of celestial objects at different wavelengths, and the spectrometer, which splits light into a spectrum so that its spectral lines can be studied. Most astronomical images are studied using computers. The pictures are broken down digitally and enhanced. This method of processing images allows astronomers to extract much more information than can be achieved by simply looking at them. Garish colours are often used to code an image and make faint details easier to see.

▷ Very large telescopes are difficult to construct: it is very hard to precisely shape a large mirror because it becomes too heavy and bows. To overcome this, the Keck telescope features a pioneering system by which the mirror is divided into thirty six hexagonal fragments, all of which can be aligned by computer-controlled mechanisms. They are held in place by precision supports. This allows the Keck telescope to sport a primary mirror which is fully ten meters in diameter, making it the largest telescope in the world. A second MMT (Multiple Mirror Telescope) is currently under construction,

▷ A simple reflecting telescope focuses light by reflecting the electromagnetic rays using mirrors of various shapes, sizes and configurations. The Newtonian design of telescope focuses light by reflecting it from a curved primary mirror, constructed in the shape of a paraboloid. The light rays are deflected by a flat secondary mirror out of the telescope tube to the side where they can be observed.

◁ Another design that uses lens and mirror is the Schmidt telescope. The collected light is focused to an inaccessible position in the tube of the telescope. To overcome this, these designs are used as cameras with a film plate placed at the focal position. The light forming an image comes to a focus on a curved plane and so the film plate also has to be curved in order to keep the whole image focused.

A much better design, called a Matsukov, corrects the image-degrading aberrations found in mirrored systems. The primary mirror is a spherical mirror; the secondary mirror is a silvered area on the spherical correcting lens. Although these telescopes produce the best images, they are impractical because of the difficulty in mounting a large lens at its edges. The largest telescopes of this design are just over a meter in diameter.

◁ Equally as important as the telescope, is the mounting upon which it is placed and the dome in which it is housed. The mounting offers a stable foundation upon which the telescope can be free from vibration. The housing provides all weather cover for the telescope as well as protection from the wind. Objects are located in the night sky by a system of co-ordinates known as right ascension and declination. The right ascension figure gives a measure of how far round the horizon the telescope should be turned and the declination shows how high up it should be pointed. The telescope will also be fitted with a drive system. This is a motor which counteracts the rotation of the Earth.

NON-OPTICAL TELESCOPES

Not all telescopes are designed to look at objects that emit light. The visible region is only a very small part of the total electromagnetic spectrum, and observing the Universe at other wavelengths opens new windows on the cosmos. However, not all of the electromagnetic radiation from these other wavelengths reaches the Earth's surface. Much of it is blocked by the Earth's atmosphere and so, in many cases, radiation has to be collected by satellites orbiting the Earth.

Telescopes in space can also perform observations continually, hindered by neither clouds nor daylight. Their chief disadvantage is that, because maintenance is so difficult, the reliability of the equipment must be very high. The Hubble Space Telescope, launched by the United States in 1990, suffered several mechanical failures that required complicated repair work to be performed by space shuttle astronauts in 1993. When the telescope is working, the collected photons of radiation are used to form an image which is then converted to electrical signals and beamed down to Earth for analysis.

The short wavelengths of electromagnetic radiation occupy the high-energy world of gamma and X rays, which need very different kinds of telescopes to detect them. They can be focused by using a grazing-incidence telescope, because gamma rays and X rays are so energetic that most photons would pass straight through a conventional mirror if it were used to reflect and focus the rays. Instead of having a parabolic shape, the mirror is extended upward and made cylindrical, so that the photons meet it at very shallow angles. The high-energy rays then gently reflect and focus. Grazing-incidence telescopes often have several cylindrical telescopes of differing radii nested within each other to improve their power.

Wavelengths longer than those of light and infrared radiation are occupied by the radio region of the spectrum. Radio waves usually reach the Earth's surface without serious distortion, and the radiation is collected by vast land-based dishes and interconnected dish systems known as interferometers. In an interferometer, two or more

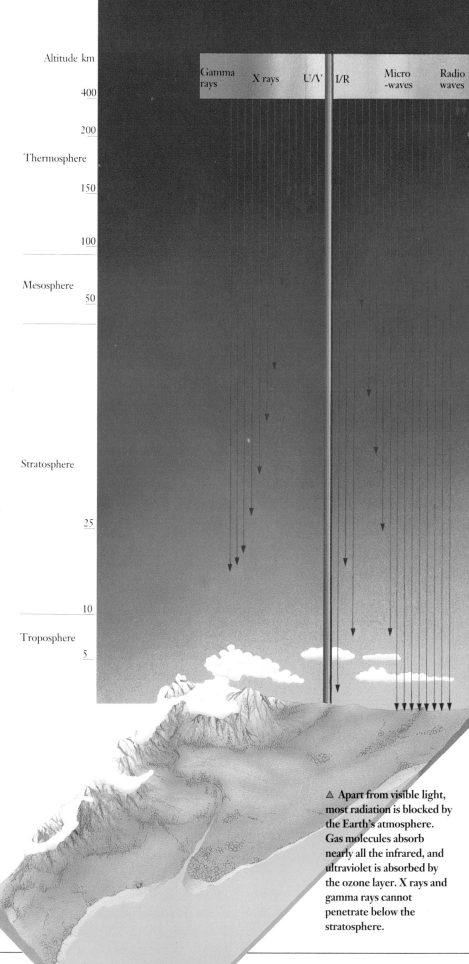

Altitude km

400

200

Thermosphere

150

100

Mesosphere

50

Stratosphere

25

10

Troposphere

5

Gamma rays | X rays | U/V | I/R | Micro-waves | Radio waves

△ Apart from visible light, most radiation is blocked by the Earth's atmosphere. Gas molecules absorb nearly all the infrared, and ultraviolet is absorbed by the ozone layer. X rays and gamma rays cannot penetrate below the stratosphere.

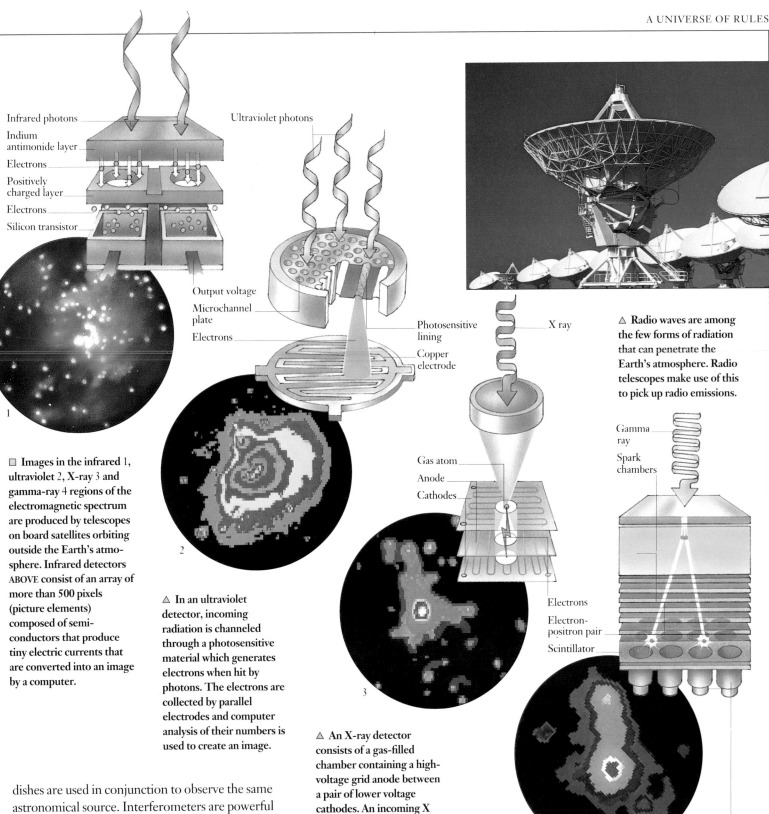

Infrared photons
Indium antimonide layer
Electrons
Positively charged layer
Electrons
Silicon transistor

Ultraviolet photons

Output voltage
Microchannel plate
Electrons

Photosensitive lining
Copper electrode

X ray

△ **Radio waves are among the few forms of radiation that can penetrate the Earth's atmosphere. Radio telescopes make use of this to pick up radio emissions.**

Gas atom
Anode
Cathodes

Gamma ray
Spark chambers

Electrons
Electron-positron pair
Scintillator

Photomultipliers

□ **Images in the infrared 1, ultraviolet 2, X-ray 3 and gamma-ray 4 regions of the electromagnetic spectrum are produced by telescopes on board satellites orbiting outside the Earth's atmosphere. Infrared detectors ABOVE consist of an array of more than 500 pixels (picture elements) composed of semiconductors that produce tiny electric currents that are converted into an image by a computer.**

△ **In an ultraviolet detector, incoming radiation is channeled through a photosensitive material which generates electrons when hit by photons. The electrons are collected by parallel electrodes and computer analysis of their numbers is used to create an image.**

△ **An X-ray detector consists of a gas-filled chamber containing a high-voltage grid anode between a pair of lower voltage cathodes. An incoming X ray knocks a high-speed electron out of a gas atom, which releases more electrons that collect on the anode. Signals from the anode to the cathodes indicate the position of the original X ray, so that an image of all incoming rays can be gradually built up.**

△ **Gamma rays are also detected using a gas-filled chamber. An incoming ray passes through a series of narrow spark chambers made of tungsten plates, and is thereby converted into an electron-positron pair. The two particles traverse wider spark**

chambers before hitting a scintillator, producing flashes of light which trigger the release of electrons, measured by a photomultiplier. This measurement denotes the energy of the original gamma ray and is used to build up an image.

dishes are used in conjunction to observe the same astronomical source. Interferometers are powerful pieces of equipment. Two radio telescopes are separated by a certain distance, called the baseline, and both are pointed toward the same astronomical source. The radio waves from the source travel slightly further to reach one telescope than the other. When the signals are combined in a computer, they do not match exactly, but produce an interference pattern. Astronomers can use this pattern to construct an image of the object they are looking at. The longer the baseline, the more detail can be seen in the image. Even more detail can be seen if more than two dishes, set out with different baselines, are used.

SPECIAL RELATIVITY

THE special theory of relativity was put forward by Albert Einstein in 1905. In it he provided the most complete mathematical description yet of the observable properties of the Universe. For the laws of physics to apply universally, those properties must be the same for any observer, regardless of whether or not he or she is stationary or in motion. The special theory provided such an explanation, although it only applies in situations where the observer's motion is constant.

If the observer is changing velocity (for example through the effect of gravity), there must be an external force acting upon him and this situation was explained by Einstein in 1915, in his general theory of relativity.

There are two guiding principles behind the special theory of relativity. The first is known as the "principle of relativity", which states that motion cannot be expressed in absolute terms, but only relative to something else. For example, if a stunt artist in a car traveling westward at 100 kilometers per hour (61 miles per hour) climbed a ladder up to an aircraft traveling at exactly the same velocity, that aircraft would seem stationary to him.

However, to an observer whose feet were planted firmly on the ground, the motion of the vehicles, and the person on the ladder, would clearly be westward at 100 km/h (61 mph). But if that same event could be watched from a point on the Sun or some other part of the Solar System well away from the gravitational influence of the Earth, the motion of the vehicles would be superimposed on the rotation of the Earth and the motion of its orbit around the Sun.

The first observer would be measuring the car's motion relative to the Earth; the second, its motion relative to the Sun. Even the Sun is not stationary, however, and if the observer were able to go beyond its gravitational influence and measure the car's motion once again, the motions of the car, ladder, aircraft, Earth and Sun would be measured relative to the nucleus of our Galaxy. And in recent years, scientists have shown that the Galaxy itself is moving through space: there are, according to the special theory of relativity, no absolutely stationary points anywhere in the Universe from which observations can be taken.

The principle of relativity also states that there is no possible experiment that can tell the person who is performing it what his or her absolute motion through space is. Climbing the ladder between car and aircraft would be just as difficult whether done on the move, or in a stationary arrangement. Only external events such as the rush of air could enable a person on the car, the aircraft or the ladder to determine whether the vehicles are moving or stationary. Similarly, the rotation of the Earth cannot be felt on Earth, and can only be observed with reference to an external event (the apparent motion of the Sun through the sky).

The second guiding postulate of the special theory of relativity is that while all other movement is relative to the point of observation, the speed of light is absolute and universal. Experiments conducted in the 1890s had shown that the speed of light remains constant, no matter how fast the experiment may be moving when performing the measurement.

Using these two ideas, Einstein found that two observers in relative motion would make different observations of length, time, speed, mass, momentum and energy, the differences increasing at higher speeds. For example, a spacecraft traveling at close to the speed of light would appear (to an external observer) to increase its mass and decrease its length. Also, time on that spacecraft would seem (to the external observer) to pass more slowly than usual. However, none of these effects would be felt on the spacecraft itself.

Two other important consequences of these two principles were shown by Einstein. The first is that nothing can travel faster than the speed of light, as its mass would become infinite at that speed. The second is that mass is simply the embodiment of energy. As the spacecraft approaches the speed of light, it increases its mass and the energy supplied to accelerate the spacecraft is transferred into mass. This equivalence of mass and energy was summed up by Einstein in his famous equation: $E=mc^2$ (energy is equal to mass times the speed of light squared). The equivalence of mass and energy is of crucial importance to our Universe, since in nuclear fusion (which occurs at the core of stars) a small amount of mass is converted into a large release of energy: and this is the energy that causes the stars to burn, and the Sun to radiate its heat to Earth and make life possible.

KEYWORDS

ABERRATION OF
    STARLIGHT
DOPPLER EFFECT
FRAME OF REFERENCE
GENERAL RELATIVITY
LENGTH CONTRACTION
MASS DILATION
PRINCIPLE OF
    EQUIVALENCE
PRINCIPLE OF RELATIVITY
RELATIVISTIC SPEED
SPACETIME CONTINUUM
TIME DILATION

□ The principle of relativity states that motion is relative to the viewpoint of an observer. A daredevil climbing from a moving car to an aircraft 1 sees the aircraft as stationary, whereas the observer on the ground 2 sees both vehicles, and the daredevil, as moving with a steady speed and direction, relative to the Earth. A theoretical observer on the Sun 3 would see the car's motion and the ground-based observer in terms of the movement of the Earth 4 as it rotates and orbits the Sun 5; whereas the observer on a star at the center of the Galaxy 6 would see the motion of the Sun around the Galaxy as well.

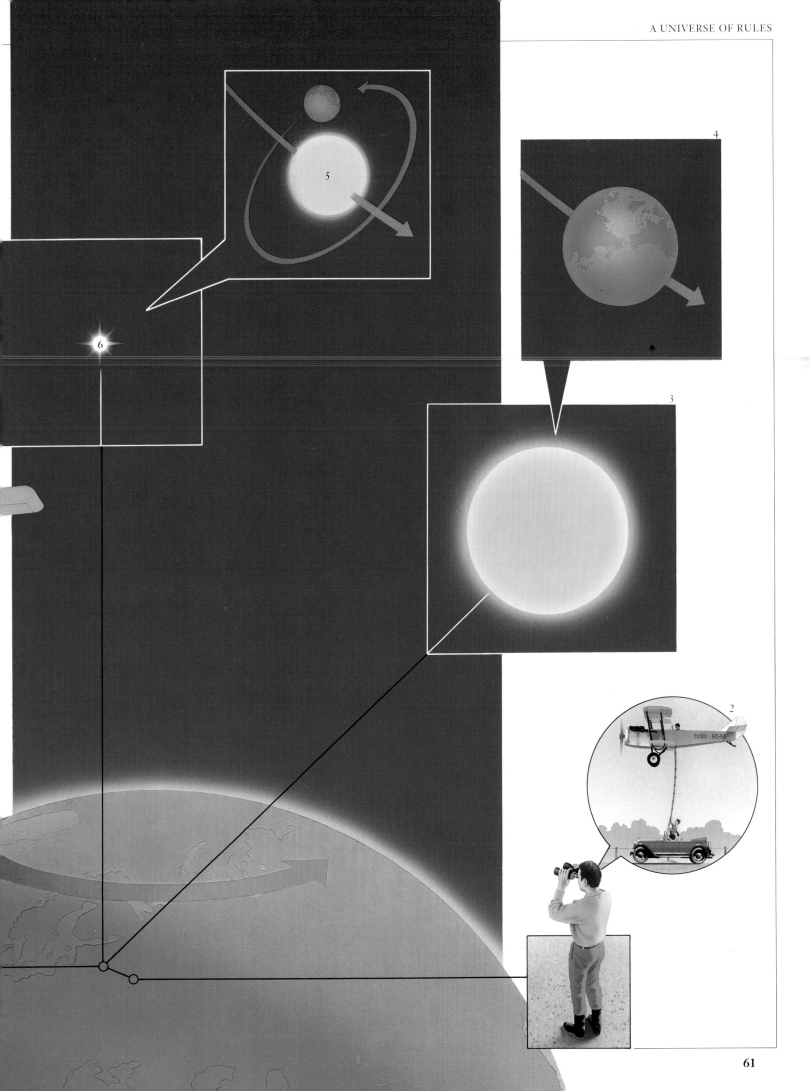

THE PARTICLE FAMILY

IN ADDITION to the five or six stable elementary particles, scientists have identified a large number of other particles. Although few of these exist for long on Earth, many of them played a role in the early history of the Universe.

Subatomic particles may be grouped together into classes of particles with similar properties. For example, all particles that exist in clearly defined structures because they obey Pauli's exclusion principle (which states that no two particles can exist in the same quantum state) are known as fermions. These include protons, neutrons and electrons. All particles that do not obey the exclusion principle are known as bosons; they include photons.

Subatomic particles may also be categorized according to how heavy they are. Leptons are lightweight particles, and there are six types. The first three are the electron, the muon and the tauon: all three have the same negative electric charge but different masses. The more massive muon and tauon can exist only for fractions of a second. The remaining leptons are the three varieties of neutrino: the electron neutrino, the muon neutrino and the tauon neutrino. These are not usually distinguished from each other and are simply referred to as neutrinos. They interact so weakly with matter that most pass straight through the Earth when they reach it. It is possible that they are dark matter – the undetected "missing" matter that may be responsible for various unexplained physical phenomena in the Universe. Leptons are affected by the weak nuclear force, whereas particles that act via the strong nuclear force are known as hadrons.

Hadrons – neutrons and protons – are heavy particles and, unlike leptons, consist of quarks. Quarks are elementary particles that come in six "flavors" known as up, down, strange, charm, truth and beauty. Truth and beauty are sometimes referred to as top and bottom. Strange and top quarks are heavier versions of up quarks, whereas charm and bottom are identical to down quarks in everything except mass. Quarks have fractional electrical charges of either positive two thirds or negative one third. They can combine in

twos or threes: protons, with positive charge of one, consist of two up quarks and a down quark; neutrons, with no electrical charge, consist of one up quark and two down quarks.

One other criterion has to be met when quarks join together. Each of the six flavors also has a property known as color. This is only an analogy; quarks are too small for the concept of color (which derives from the wavelength of visible light) to have any meaning. Quarks are said to be either red, green or blue, and when they join together in triplets there must be a quark of each color. Such quark triplets are known as baryons. A particle made of only two quarks is known as a meson. In order to make a particle of two quarks and satisfy the color requirement, a different kind of matter, known as antimatter is needed.

According to Einstein's equation $E = mc^2$, mass is the embodiment of energy. If enough energy is present in a region of space to convert the energy into a particle of matter, a mirror-image counterpart is also created at the same time. When these two particles, or ones like them, come into contact they annihilate each other and return the energy bound up in their mass to the Universe. These mirror-image particles are known as antimatter. For example, the antimatter

△ Many subatomic particles exist only for fleeting instants in high energy states, and can be detected only in high-energy collisions – or in cosmic rays, as shown here. A sulfur nucleus (red) has collided with a nucleus in photographic emulsion used to capture the rays. The particles produced by the collision include a fluorine nucleus (green), pions (yellow) and nuclear fragments (blue).

Quark flavors
- Up
- Down
- Strange
- Charm
- Bottom
- Top
- Antiup
- Antidown
- Antistrange
- Anticharm
- Antibottom
- Antitop

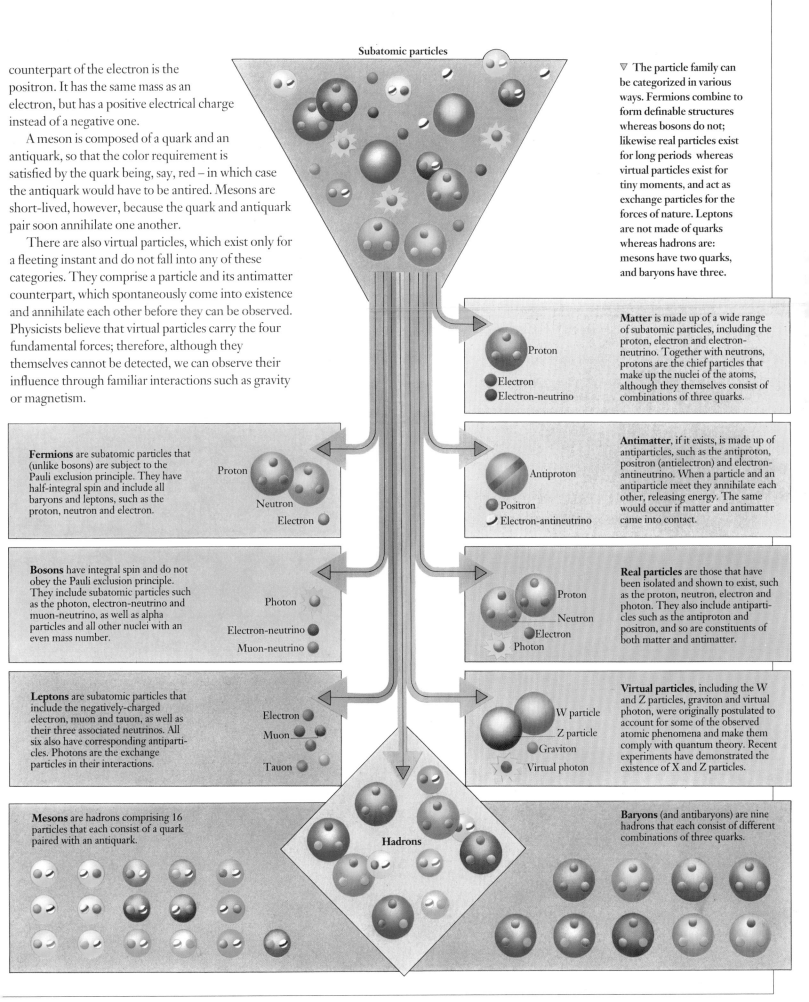

Subatomic particles

counterpart of the electron is the positron. It has the same mass as an electron, but has a positive electrical charge instead of a negative one.

A meson is composed of a quark and an antiquark, so that the color requirement is satisfied by the quark being, say, red – in which case the antiquark would have to be antired. Mesons are short-lived, however, because the quark and antiquark pair soon annihilate one another.

There are also virtual particles, which exist only for a fleeting instant and do not fall into any of these categories. They comprise a particle and its antimatter counterpart, which spontaneously come into existence and annihilate each other before they can be observed. Physicists believe that virtual particles carry the four fundamental forces; therefore, although they themselves cannot be detected, we can observe their influence through familiar interactions such as gravity or magnetism.

▽ The particle family can be categorized in various ways. Fermions combine to form definable structures whereas bosons do not; likewise real particles exist for long periods whereas virtual particles exist for tiny moments, and act as exchange particles for the forces of nature. Leptons are not made of quarks whereas hadrons are: mesons have two quarks, and baryons have three.

Matter is made up of a wide range of subatomic particles, including the proton, electron and electron-neutrino. Together with neutrons, protons are the chief particles that make up the nuclei of the atoms, although they themselves consist of combinations of three quarks.

Proton
Electron
Electron-neutrino

Fermions are subatomic particles that (unlike bosons) are subject to the Pauli exclusion principle. They have half-integral spin and include all baryons and leptons, such as the proton, neutron and electron.

Proton
Neutron
Electron

Antimatter, if it exists, is made up of antiparticles, such as the antiproton, positron (antielectron) and electron-antineutrino. When a particle and an antiparticle meet they annihilate each other, releasing energy. The same would occur if matter and antimatter came into contact.

Antiproton
Positron
Electron-antineutrino

Bosons have integral spin and do not obey the Pauli exclusion principle. They include subatomic particles such as the photon, electron-neutrino and muon-neutrino, as well as alpha particles and all other nuclei with an even mass number.

Photon
Electron-neutrino
Muon-neutrino

Real particles are those that have been isolated and shown to exist, such as the proton, neutron, electron and photon. They also include antiparticles such as the antiproton and positron, and so are constituents of both matter and antimatter.

Proton
Neutron
Electron
Photon

Leptons are subatomic particles that include the negatively-charged electron, muon and tauon, as well as their three associated neutrinos. All six also have corresponding antiparticles. Photons are the exchange particles in their interactions.

Electron
Muon
Tauon

Virtual particles, including the W and Z particles, graviton and virtual photon, were originally postulated to account for some of the observed atomic phenomena and make them comply with quantum theory. Recent experiments have demonstrated the existence of X and Z particles.

W particle
Z particle
Graviton
Virtual photon

Mesons are hadrons comprising 16 particles that each consist of a quark paired with an antiquark.

Hadrons

Baryons (and antibaryons) are nine hadrons that each consist of different combinations of three quarks.

THE QUANTUM VIEW

ELECTROMAGNETIC radiation is often presented as a wave (that is, as energy), whereas atoms and subatomic particles have been described as solid objects (matter). However, the Universe is not as precise as this in its definitions of matter and energy, and the true nature of its components is much more ambiguous. Because it is not possible to visualize the true nature of atomic structure or the nature of energy propagation, we must rely on models. Models are a means by which humans can think of scientific concepts in order to understand them better. For example, when considering gases, the constituent atoms can be thought of as billiard balls colliding with one another. When thinking of conductivity in semi-conductors, however, a more complex model of atoms is required.

Some phenomena involving light, such as atomic absorption and emission, can be understood only if light is thought of as consisting of a stream of particles. These particles are called photons. Yet light rays combine with one another in a manner that is understandable only if light is a wave. If that is the case, what is light: waves or particles? The best answer is that light is both a wave and a particle – and it is also neither. The true nature of light must be something very complex, and at present it is beyond human comprehension.

For this reason, scientists have to rely on two apparently contradictory models of the nature of light. The trick is in thinking of light as a particle to solve one problem and then thinking of it as a wave to solve another. This way of thinking is known as wave-particle duality.

With the successful application of the idea that the energy carried in light rays is transported in discrete wave-packets, the concept of quantization was born. This idea, that energy can only be transported and exchanged in well-defined quantities, is the foundation of modern quantum theory, which was formulated early in the 20th century based on the work of several scientists – notably Max Planck, Erwin Schrödinger, Werner Heisenberg and Niels Bohr. The phenomenon of quantization occurs on such a small scale that it goes unnoticed in the everyday world; it can only be observed on the scale of subatomic particles.

△ The Danish physicist Niels Bohr explained the spectral lines emitted by hydrogen by defining certain orbits for the electrons around the nucleus, corresponding to the only positions in which the electrons can exist around it. Puzzling over whether particles could be wavelike (because waves could be particle-like), Louis de Broglie derived an equation for the wavelength of a particle. It depends on the particle's mass and velocity. He also showed that the circumference of the orbits around hydrogen nuclei match a whole number of de Broglie wavelengths for the orbiting electrons.

▷ In this experiment, electrons can be shown to behave like particles. They are emitted at the negatively charged cathode and accelerated by a positively charged anode. As they fly along the vacuum tube, they hit a metal flag which can be turned so that the angles at which the electrons strike vary. As a result, the electrons to bounce off in different directions depending upon the angle at which they strike the flag. The electrons can be detected on a screen similar to that of a television.

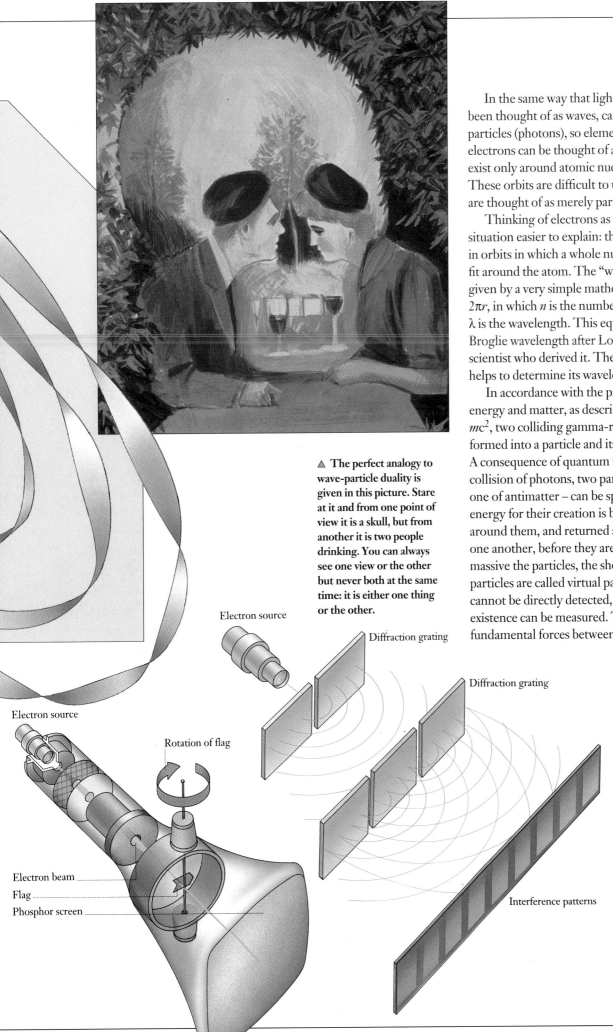

In the same way that light, which had traditionally been thought of as waves, can be thought of as particles (photons), so elementary particles such as electrons can be thought of as waves. Electrons can exist only around atomic nuclei in certain orbits. These orbits are difficult to understand if the electrons are thought of as merely particles.

Thinking of electrons as wave functions makes the situation easier to explain: the electrons can exist only in orbits in which a whole number of wavelengths can fit around the atom. The "wavelength" of a particle is given by a very simple mathematical expression, $n\lambda = 2\pi r$, in which n is the number of waves in an orbit, and λ is the wavelength. This equation is known as the de Broglie wavelength after Louis de Broglie, the French scientist who derived it. The energy of the electron helps to determine its wavelength.

In accordance with the principle of equivalence of energy and matter, as described by the equation $E = mc^2$, two colliding gamma-ray photons can be transformed into a particle and its antimatter counterpart. A consequence of quantum theory is that, without a collision of photons, two particles – one of matter and one of antimatter – can be spontaneously created. The energy for their creation is borrowed from the space around them, and returned as the particles annihilate one another, before they are detected. The more massive the particles, the shorter their lifespan. These particles are called virtual particles. Although they cannot be directly detected, the effects of their existence can be measured. They carry the fundamental forces between elementary particles.

△ The perfect analogy to wave-particle duality is given in this picture. Stare at it and from one point of view it is a skull, but from another it is two people drinking. You can always see one view or the other but never both at the same time: it is either one thing or the other.

Electron source

Diffraction grating

Diffraction grating

Electron source

Rotation of flag

Electron beam

Flag

Phosphor screen

Interference patterns

◁ The wavelike nature of electrons can be highlighted in diffraction experiments. The electrons are passed through two slits approximately the size of the de Broglie wavelength of the electron. This causes them to diffract like light rays through a grating. When the two wavefronts come into contact again they cause interference to occur, and light and dark patches appear on the viewing screen. This echoes a classic experiment performed by the British physicist Thomas Young around 1800 in which he used two slits and a beam of light to demonstrate the wavelike nature of light.

EXCLUSION AND UNCERTAINTY

ELECTRONS around atoms are thought of as existing in quantum states rather than being in orbits. This is because each electron can have its quantum state uniquely defined by the properties known as energy, angular momentum and spin. The electrons congregate into groupings, known as electron "shells", according to the energy they possess. Within these electron shells they are grouped into subshells according to their angular momentum.

Finally, electrons can also have spin. This indicates the direction of the electron's magnetic field.

No two electrons around any atom can exist within the same quantum state; to do so would be like two objects trying to occupy the same physical space on a table top. For example, if two electrons, orbiting a particular nucleus, have the same energy and angular momentum, then their spins must be different. Then they are not in the same quantum state.

This exclusion of electrons from certain states is known as Pauli's exclusion principle (after the Austrian-Swiss physicist Wolfgang Pauli [1900–58]). Because electrons are restricted to certain quantum states, this imposes a very well-defined structure on atoms. This structure is what governs the way in which the atoms behave. In short, it gives rise to most of the physical phenomena in the Universe.

▷ The time for which virtual particles can exist depends upon their mass: the more massive they are, the less time they can exist, because their lifetime is calculated by dividing the Planck constant by their mass. If something with the same mass as a coin could exist for one second, then a helium atom could exist for ten million years. A single proton, on this scale, would persist for 100 million years. A human could last for one hundred-thousandth of a second and a car for one millionth of a second.

◻ The uncertainty principle has interesting consequences for our perception of a vacuum. A cathode-ray tube, such as that in a television set or computer monitor, providing it is turned off, contains a vacuum. According the classical view BELOW LEFT, that vacuum will simply be empty space. In practice this is impossible, and there will always be a handful of atoms present. In the quantum view BELOW RIGHT, a consequence of Heisenberg's uncertainty principle is that virtual particles can exist in a vacuum for fractions of a second which we cannot measure. We are therefore unaware of their existence.

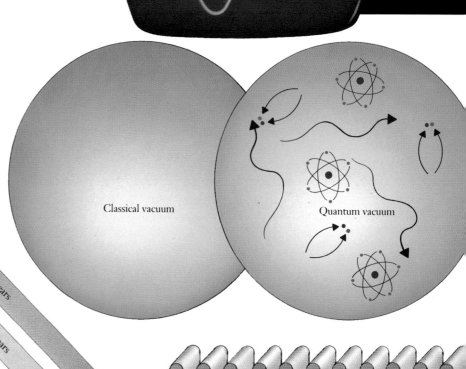

Proton

Atom

10^8 years

Coin

10^7 years

Human

1 second

Car

10^{-5} sec

10^{-6} sec

Classical vacuum

Quantum vacuum

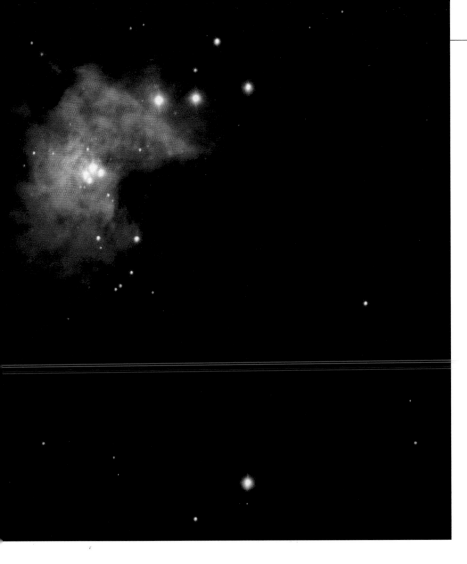

◁ In this view of the star-forming region M42 in the constellation of Orion, the red glow comes from atoms of oxygen which have had two of the outer electrons stripped from them by the intense ultraviolet radiation produced by the central stars. The single wavelength of light produced by these atoms is known as a forbidden line, because the electrons cannot reach these orbital states except under the conditions present in space. It is very hard to reproduce these conditions in a laboratory, and for many years astronomers wondered how such light was produced.

The exclusion principle gives firm rules about electrons around atomic nuclei, but another important principle establishes uncertainty about their position and momentum. The uncertainty principle is based on wave-particle duality: talking about a particle implies a definite position in space, whereas a wave is usually thought of as an an extended object stretching across space. Using simple mathematics, the German physicist Werner Heisenberg (1901–76) showed that it is possible to locate a portion of a wave, called a wave packet, which could then be thought of as a particle: a photon. Another example is the de Broglie representation of an electron. However, the localization is only possible to a certain level of precision. It is impossible to know exactly where a wave packet or, indeed, a particle is and also know exactly in which direction it is heading. The more precisely a particle's position is measured, the less the details of its motion can be known. Particles therefore remain slightly mysterious.

The uncertainty principle is a fundamental property of the Universe on its smallest scale. The quantities of time and energy are also linked by the uncertainty principle. Particles that embody energy as mass can live for a certain time, provided that the lifetime multiplied by the amount of energy does not exceed the Planck constant. It is this aspect of the uncertainty principle that explains how virtual particles can come into existence fleetingly and then disappear again.

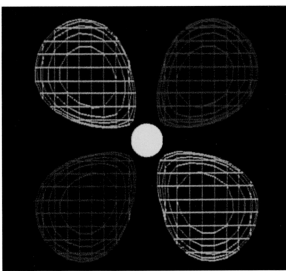

◁ The positions of electrons around an atom are quantified by quantum numbers. Each electron shell contains atoms of the same energy. The shells can be divided into subshells, which group the electrons according to the angular momentum of their orbits. The orientation of an electron depends upon the initial shell and the subshell in which it is located. Orbital diagrams show the regions in which electrons of the same subshells are found.

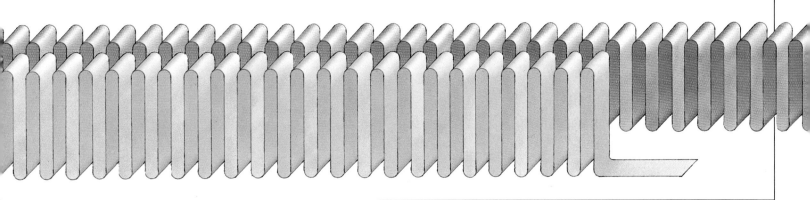

FORCES AND FIELDS

THE four fundamental forces of nature – gravity, electromagnetism, the strong nuclear force and the weak nuclear force – apply to every particle of matter in the Universe. In fact, it is through the fundamental forces that separate pieces of matter "communicate" with each other. The four forces are of different strengths and apply on different scales: planets orbit the Sun because of gravity, but electrons orbit atomic nuclei because of electromagnetism. Each of the forces of nature is carried by a different kind of virtual particle.

Of the four fundamental forces, we are most familiar with gravity and electromagnetism, because their effects are most obvious in the world around us. The other two forces act only within atomic nuclei, so they are less noticeable.

Gravity is the natural force of attraction that acts between objects with mass. The greater the mass of two objects, the greater the force of their mutual attraction. The farther the objects are from each other, however, the weaker is the force of gravity between them. This is because gravity follows an inverse-square law: if the distance between the two objects is doubled, the force between them is quartered. Gravity is the weakest of the fundamental forces, yet it has an unlimited range. It shapes the Universe on its largest and most dramatic scales, because it acts over such vast distances throughout space.

The electromagnetic force acts between all particles with an electric charge, such as electrons, protons and ions. It is the driving force in all chemical reactions, which rely on interactions between electrons in order to to form molecules. The force consists of two interconnected forces, electricity and magnetism. A moving particle with an electric charge creates a magnetic field, whereas a magnetic field around a conductive substance induces charged particles to move.

The electromagnetic force is different from gravity in that, as well as an attractive, it also has a repulsive element. These are characterized by assigning positive and negative signs to

▷ Electromagnetism binds atoms as opposite charges attract. Lightning occurs when the ground and the clouds are at different electrical potentials. To equalize this, charges (electrons) move from the negative region to the positive region.

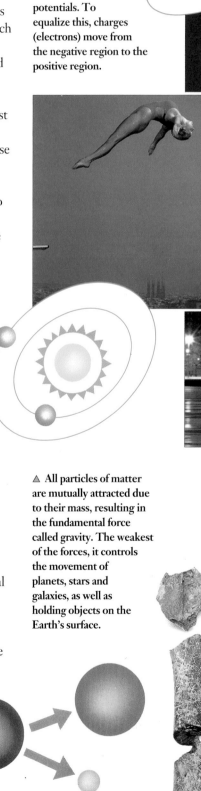

▲ All particles of matter are mutually attracted due to their mass, resulting in the fundamental force called gravity. The weakest of the forces, it controls the movement of planets, stars and galaxies, as well as holding objects on the Earth's surface.

charges to show their polarity. Unlike charges attract, whereas like charges repel. This is why negatively-charged electrons remain in orbit around positively-charged atomic nuclei. The electromagnetic force is similar to gravity, however, in that it, too, follows an inverse-square law. Although it is stronger than gravity, it does not dominate the structure of the Universe because, over large volumes, any overall positively- or negatively-charged regions cancel out each other.

The nuclear forces are extremely strong, but they are confined to the nuclei of atoms. The strong nuclear force – the strongest of the fundamental forces – acts only over a distance comparable to the diameter of a proton or a neutron, about 10^{-15}m. It holds protons and neutrons together to form atomic nuclei. This is the force that must be overcome in nuclear fission in order to "split" the atom.

The weak nuclear force has an even smaller range than the strong nuclear force – only about 10^{-18}m, the diameter of an electron. Within its range, it is stronger than gravity but not quite as strong as electromagnetism. It governs the creation and interaction of the elementary particles known as neutrinos. These are created when neutrons become protons or protons become neutrons. Neutrinos interact very weakly with atomic matter because it is necessary for them almost to touch the nucleus before the weak force can cause them to interact.

◁ The weak nuclear force is stronger than gravity but weaker than either the strong nuclear force or electromagnetism. It governs the radioactive decay of some atoms, becoming active every time a hadron (a particle composed of quarks, such as a proton) turns into a new hadron (such as a neutron) and a lepton (such as an electron, which is not composed of quarks). Some methods of radioactive dating rely on this decay process. Every time such a reaction takes place, a neutrino is released or absorbed.

◁ The strong nuclear force, which holds together the nucleus of an atom, must be overcome to produce a nuclear explosion. When this happens, vast amounts of energy are liberated. Nuclear reactors do the same but in a controlled way, and use the heat to drive turbines and generate electricity.

Unifying the Forces

Although the four fundamental forces are very different from each other, it is possible that they are simply aspects of a single superforce. For several decades scientists have been looking for experimental evidence that will support such a possibility. A substantial step toward the discovery of this superforce would be a theory that united both nuclear forces with the electromagnetic force. Such a theory is known as a Grand Unified Theory, or GUT; many versions have been put forward although none has yet proved better than the others. Although the details of the various GUT theories are different, their general aims are the same.

An important first step was achieved when it was demonstrated, in a particle accelerator, that at sufficiently high temperatures it becomes impossible to tell the difference between an electron and a neutrino. The two particles are usually differentiated by their effect on other particles. Because electrons interact using the electromagnetic force and neutrinos interact via the weak nuclear force, the way they behave with other particles is usually very different. Electromagnetism is carried by virtual photons, whereas the weak nuclear force is carried by virtual particles known as the W particle and the Z particle. At high energies both particles interact in exactly the same way, using an amalgamation of the two individual forces. This combined force is known as the electroweak force.

In the final stage in a Grand Unified Theory, quarks behave in the same way as leptons. This is projected to happen at particle energies corresponding to enormous temperatures of 10^{27} K. This is the temperature that the Universe is thought to have had when it was only 10^{-35} seconds only. This aspect of the GUT is likely to remain theoretical, because it is almost inconceivable that scientists could build a machine to reproduce this temperature.

The GUT theory would remain untested if not for one tiny effect that may still be observable in the low-energy Universe of today. GUTs predict that protons should decay into lighter particles via a strong nuclear force interaction. This decay is very slow: the average lifetime of a proton is calculated to be 10^{32} years. But, given a large enough assemblage of protons – such as in a huge tank of water – the statistical chances of

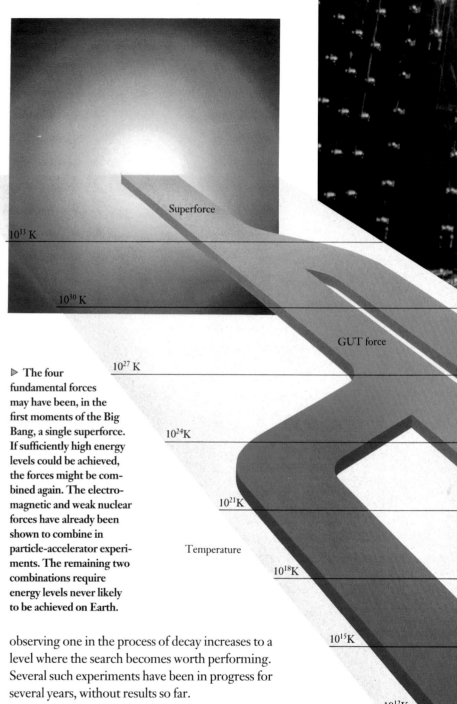

▷ **The four fundamental forces may have been, in the first moments of the Big Bang, a single superforce. If sufficiently high energy levels could be achieved, the forces might be combined again. The electromagnetic and weak nuclear forces have already been shown to combine in particle-accelerator experiments. The remaining two combinations require energy levels never likely to be achieved on Earth.**

Superforce

10^{33} K

10^{30} K

10^{27} K

10^{24}K

10^{21}K

10^{18}K

10^{15}K

10^{12}K

10^{9}K

GUT force

Temperature

observing one in the process of decay increases to a level where the search becomes worth performing. Several such experiments have been in progress for several years, without results so far.

Following the completion of a Grand Unified Theory, the task of unifying gravity with the other three forces could be undertaken. But before this can happen, modern physics must develop a quantum theory of gravity – one that requires a graviton to carry the force between masses. This particle would carry gravity in the same way that photons carry the electromagnetic force. Such gravitons are proving difficult to develop theories for, however. The best theory for gravity was developed by Albert Einstein and is known as the general theory of relativity. It is not a quantum theory because it does not require a virtual particle to carry the force.

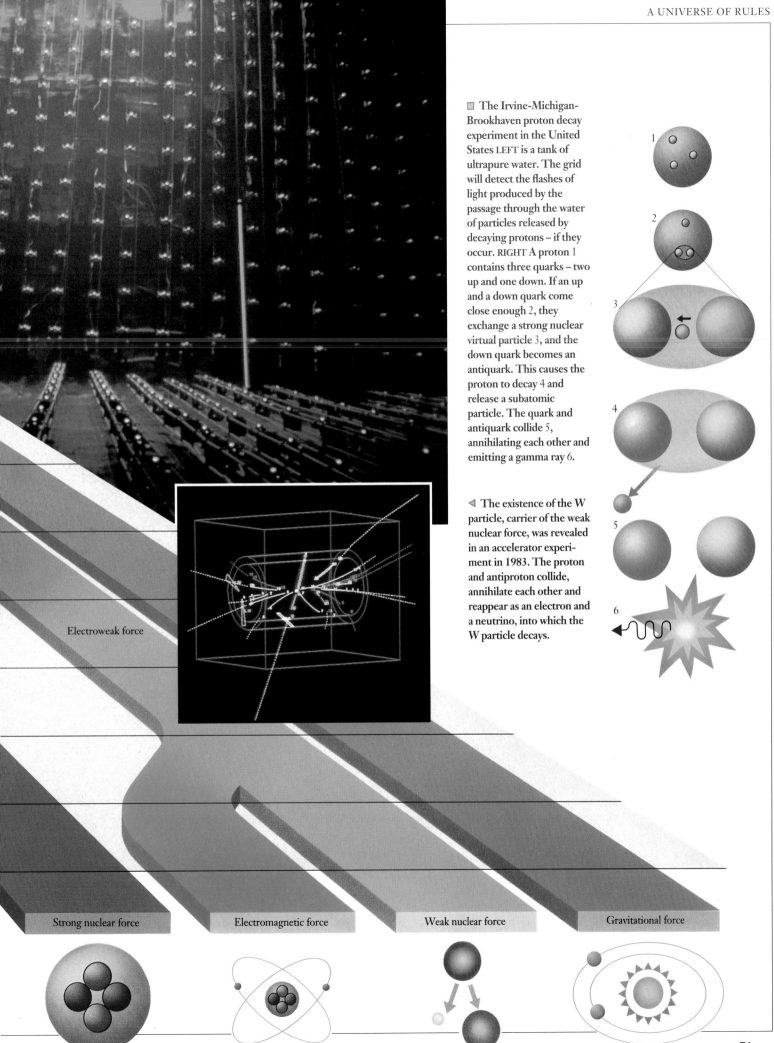

□ The Irvine-Michigan-Brookhaven proton decay experiment in the United States LEFT is a tank of ultrapure water. The grid will detect the flashes of light produced by the passage through the water of particles released by decaying protons – if they occur. RIGHT A proton 1 contains three quarks – two up and one down. If an up and a down quark come close enough 2, they exchange a strong nuclear virtual particle 3, and the down quark becomes an antiquark. This causes the proton to decay 4 and release a subatomic particle. The quark and antiquark collide 5, annihilating each other and emitting a gamma ray 6.

◁ The existence of the W particle, carrier of the weak nuclear force, was revealed in an accelerator experiment in 1983. The proton and antiproton collide, annihilate each other and reappear as an electron and a neutrino, into which the W particle decays.

Electroweak force

Strong nuclear force

Electromagnetic force

Weak nuclear force

Gravitational force

GENERAL RELATIVITY

THE SPECIAL theory of relativity, developed in the first years of the 20th century by Albert Einstein, described observers and systems in uniform constant motion relative to each other. Einstein wanted to extend this work to cases in which a system is changing its velocity, as when something is accelerating. By extending relativity to accelerating frames of reference, Einstein also formulated a new theory of gravity. So far, it has proved to be the most valid.

The cornerstone of the general theory of relativity is the principle of equivalence. This principle states that the conditions present in a potential well – the region around an object with a gravitational field – can be exactly reproduced by an accelerating frame of reference. A frame of reference with a force acting upon it – such as gravity – can be counteracted by the application of a correctly chosen applied acceleration. Thus, by implication, forces and accelerations are equivalent.

Using the laws of general relativity, the three familiar dimensions of space – up and down, left and right, and in and out – can be linked with the one dimension of time, forward. They can be thought of as a four-dimensional spacetime continuum. Anything that

moves through the Universe travels along straight lines, called geodesics, within this continuum. Geodesics are are straight, but the continuum is often curved. This curvature occurs when massive objects such as stars, planets and, on a much larger scale, galaxies distort the spacetime continuum into potential wells. Photons of electromagnetic radiation follow straight paths on th spacetime continuum, but when they venture close to a potential well, these translate into curved lines in three-dimensional space. One result is the phenomenon known as gravitational lensing, by which a far-distant celestial object, such as a quasar, is split into two or more images, because of the gravitational field of an intervening galaxy.

▷ **According to general relativity, the Sun should cause deformation in the spacetime continuum, and delay radio signals which have to pass close to the Sun. These effects were tested by NASA when it sent its Viking space probes to Mars in the mid-1970s. When Mars was on the far side of the Sun, NASA timed the journey of radio signals. They appeared to take 100 microseconds too long to reach the Earth. The extra time, the equivalent of the radio waves travelling an extra 30 km, was interpreted as the radio waves dipping into and back out of the gravitational well of the Sun.**

General relativity explains the effects of gravity by asking us to imagine that it is no longer a force. This can be understood by the analogy of traveling in a car going around a bend in the road. The passengers are pulled to the side by a force known as centrifugal force. But this is only an apparent force. The passenger's bodies actually try to continue in a straight line but come into contact with the side of the car which is now traveling in a new direction. Gravity can be thought of as an apparent force somewhat like centrifugal force.

The concept of straight geodesics on a curved spacetime continuum can be visualized by analogy with two people at the Earth's surface. They stand on the Equator but at different longitudes. Both face due north and begin walking. Neither would doubt that they have started walking on parallel paths and parallel lines do not converge. If they continue to walk towards the North Pole in straight lines, they get closer together. If they walk at the same speed they will meet at the North Pole. It seems as if an attractive force has pulled them together, but all they have done is follow straight paths on a curved surface: the Earth.

Gravity can be thought of in this way, but because humans are three-dimensional beings we cannot perceive the curvature of the Universe through a fourth dimension. The effect of this curvature is the force that is called gravity.

▽ **The shortest route from Europe to North America appears to be a straight line on a two-dimensional map of the Earth's surface. The world is three-dimensional, however, so the actual route is a curve. This is similar to the way objects and radiation move through the spacetime continuum. Although they appear to travel through three dimensions, they actually move along curved lines in four dimensions.**

2

THE
Big Bang

THE UNIVERSE is expanding, a fact discovered by the American astronomer Edwin Hubble in the 1920s. When studying the spectral lines characteristic of particular elements in the light emitted by distant galaxies, Hubble found that every line had moved toward the red, longer wavelength, end of the spectrum, implying that the light waves had been stretched.

This might mean that the galaxies are all moving away from us. In fact, however, the galaxies do not move, but the spacetime continuum itself is expanding, although this expansion may be overcome locally by force of gravity. Space does not expand on the Earth, within the Solar System, or even within a galaxy, but between groups of galaxies space does expand. The galaxies are driven apart just as currants in a cake mixture are moved apart as the dough rises.

As the Universe expands, the light waves from these galaxies are stretched, shifting them toward the red end of the spectrum. The farthest ones shift the most. The phenomenon is called red shift.

Presumably, therefore, the Universe used to be smaller, and may once have been infinitely small. The logic led to the theory of the Big Bang, the initial event in which the Universe and everything in it – space, time, matter, energy, even the laws of physics and the fundamental forces of nature – were created.

Visualizing the creation of the Universe is impossible. Despite the familiar name given to this event, it was neither big (space was infinitely small immediately after the moment of creation), nor was there a bang. All the matter that was, everything that is and everything that ever will be was created in that briefest of instants. Since then the Universe has expanded, cooled and become less dense, allowing galaxies, stars and planets to form. The study of the Big Bang is part of the field known as cosmology.

SCALE OF THE UNIVERSE

ASTROPHYSICS encompasses every conceivable scale that the Universe has to offer. Some of these scales appear very different from the ones with which we are most familiar – the scales of millimeters to thousands of kilometers. Beyond this limited range, our imaginations have to work harder. The Universe looks very different on these different scales, but the laws of physics apply at all of them.

On the smallest scales accessible to modern science – about 10^{-16}m – matter is made of fundamental particles known as quarks. They combine into triplets to form the elementary particles protons and neutrons. Most of the mass of an atom is concentrated into its nucleus, which has a diameter of 10^{-13}m. But virtually all the atom's volume is taken up by electrons. These exist around the nucleus in a region often referred to as the electron cloud. The electron cloud is approximately 1000 times the diameter of the nucleus, or 10^{-10}m.

The human scale lacks both the quantum phenomena of the subatomic scales and the large-scale consequences of relativity. We can look through a magnifying glass and remain unaware of the quantum interactions that cause the photons to reflect off the object, strike our eyes and allow us to view a small object on a larger scale. On still larger scales we measure in tens, hundreds or thousands of meters. These may be conveniently expressed by exponents. The diameter of the Earth, for example, is 10^7m (on a scale of 10,000 kilometers). The distance between the Earth and the Sun is about 149 million kilometers, or one Astronomical Unit (AU). Both are part of the Solar System. Mercury, the planet nearest to the Sun, has a mean distance from Earth of 0.39 AU; the farthest, Pluto, is at a mean distance of 39.44 AU.

When there are too many kilometers or astronomical units for the human mind to grasp, astronomers measure in light-years. One light-year is the equivalent of 95 trillion kilometers or 63,240 AU. The outer regions of the Solar System, called the Oort cloud, may stretch about a quarter of the way to Proxima Centauri, the nearest star – 4.3 light-years away. A rocket traveling at 10 kilometers per second (36,000 kilometers per hour) would take 100,000 years to reach this "nearby" star.

KEYWORDS

ATOM
BARYON
CLUSTERS OF GALAXIES
LIGHT YEAR
LOCAL GROUP
LOCAL SUPERCLUSTER
MATTER
MILKY WAY
OORT CLOUD
QUARK
SOLAR SYSTEM
UNIVERSE

▷ The visible Universe is defined by its age: the Universe is 15 billion years old, and it is not possible to see objects more than 15 billion light-years away. A vast number of galaxies can be detected within this limit, and some astronomers believe that equal numbers exist that are forever beyond our reach.

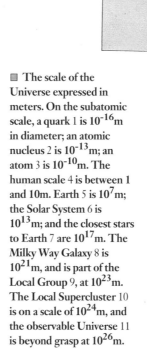

■ The scale of the Universe expressed in meters. On the subatomic scale, a quark 1 is 10^{-16}m in diameter; an atomic nucleus 2 is 10^{-13}m; an atom 3 is 10^{-10}m. The human scale 4 is between 1 and 10m. Earth 5 is 10^7m; the Solar System 6 is 10^{13}m; and the closest stars to Earth 7 are 10^{17}m. The Milky Way Galaxy 8 is 10^{21}m, and is part of the Local Group 9, at 10^{23}m. The Local Supercluster 10 is on a scale of 10^{24}m, and the observable Universe 11 is beyond grasp at 10^{26}m.

The Solar System exists in one of the spiral arms of the Milky Way galaxy – a vast system containing more than 100 billion stars in an area of 80,000 to 100,000 light-years in diameter, with the Sun at a distance of about 28,000 light-years from the center. Every star visible in the night sky is part of the Milky Way.

The Milky Way is part of a cluster known as the Local Group, with a radius of about 2.5 million light-years. Its nearest neighbor in the Local Group is 160,000 light-years away. The Andromeda galaxy, at 2.3 million light-years, is the most distant object visible to the naked eye under good conditions. The Local Group belongs to the Local Supercluster, which has a radius of 50 million light-years.

NATURAL HISTORY OF THE BIG BANG

Astronomers believe that the Universe, including matter and space, was created in a Big Bang, and that the essential processes occurred in the first tiny fractions of a second after that Bang, when temperatures were vastly higher than in the Universe today.

People often ask what existed before the Big Bang and what the Universe expanded into. Yet the concept "before the Big Bang" has no meaning, because time itself did not exist until it was created in the Big Bang. And if space, like time, was created during the Big Bang, and if space itself is expanding, then it need not be expanding into anything.

The Universe has been evolving ever since the moment of creation, and theoretical physicists and cosmologists have provided a description of the probable sequence of events that gave rise to the Universe as we know it.

In the very first instants, up to 10^{-43} seconds after the Big Bang, space and time were still forming. The forces of nature were combined into a single, primordial superforce. This is known as the Planck time, and its details can never be explained, since the laws of physics were still being defined.

By 10^{-35} seconds, space had expanded sufficiently for temperatures to have fallen to 10^{27} K, carried by extremely energetic photons. Gravity had already become a separate force, and the grand unified theory (GUT) force now separated into the strong nuclear and the electroweak forces, accompanied by the rapid creation of quarks, leptons and their antimatter counterparts. This process caused the Universe to undergo a short but huge inflation (lasting 10^{-32} seconds) before resuming its previous rate of expansion.

At 10^{-12} seconds, the electroweak force separated to form electromagnetism and the weak nuclear force, so that all four forces of nature were now separate and distinct. The particles in the Universe and their antiparticles were in a constant state of creation and annihilation, and leptons separated into neutrinos and electrons. Quarks still existed as separate entities since the temperature of the Universe precluded their joining to form heavier particles.

By 10^{-6} seconds, quarks combined into pairs or triplets, forming mesons and baryons (including protons and neutrons): since that moment quarks have been unable to exist independently. Their antiparticles did the same and then annihilated with the matter, but a tiny residue (one particle in every billion) was left which went on to form all the matter in the present-day Universe. A large number of photons also resulted from this process.

At the end of the first second in time, the temperature had fallen to 10^{10} K; five seconds later neutrinos and antineutrinos had ceased to interact with the other forms of matter. After the Universe had reached the age of 10 seconds, protons and neutrons began to come together to form deuterium nuclei.

Between one and five minutes, the strong nuclear force took hold, bringing neutrons and protons together in helium nuclei, preventing the neutrons from decaying into protons and electrons. The relative proportions of hydrogen to helium in the Universe was defined during this time. Energy levels were still so high that the atoms were entirely ionized and existed as atomic nuclei in a sea of electrons.

Some 300,000 years after the Big Bang, temperatures had fallen sufficiently – to about 3000 K – so that the electromagnetic force allowed electrons to be captured by the atomic nuclei. As space was no longer filled with a sea of stray electrons, photons could for the first time travel large distances without interacting with matter: the Universe became

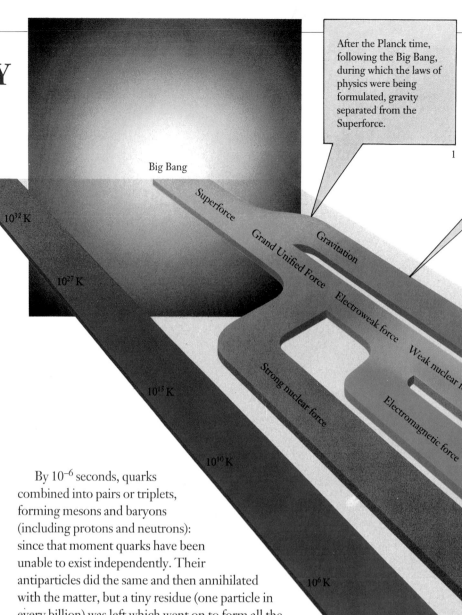

After the Planck time, following the Big Bang, during which the laws of physics were being formulated, gravity separated from the Superforce.

Big Bang

Superforce

Grand Unified Force

Gravitation

Electroweak force

Weak nuclear force

Strong nuclear force

Electromagnetic force

10^{32} K
10^{27} K
10^{15} K
10^{10} K
10^{6} K
3000 K

Temperature

▲ The current average temperature of the Universe is 3 K (detectable today as the cosmic background radiation) but it was originally much hotter. At the end of the Planck era, its temperature was 10^{32} K. Energy was carried by photons, but the early Universe was so dense that these could not travel far before being reabsorbed. Temperatures have dropped ever since.

The next key event was the separation of the electroweak and strong nuclear forces. The Universe expanded 10^{50} times in only 10^{-32} seconds.

■ Before 10^{-43} seconds, nothing can be said about conditions in the early Universe 1, but by 10^{-35} seconds, two forces of nature were distinct and the lightest particles, quarks and leptons, had come into existence 2.

By 10^{-12} seconds 3 all the particles were in a constant state of creation and annihilation; it was not until 10^{-6} seconds 4 that quarks came together to form neutrons and protons, although almost all of these also were annihilated in collisions with their antiparticles. The residue formed the matter that we now find in the Universe 5. Much later, 15 seconds after the Bang, these protons and neutrons joined together to form deuterium nuclei 6, and after a few minutes helium nuclei (two protons and two neutrons) had been created 7. After 300,000 years, atoms were formed as electrons were captured by the nuclei 8, and gravity, the weakest of the four forces, began to shape the Universe, causing matter to collect in clouds that later formed galaxies and stars.

- ⬤ Proton
- ◑ Antiproton
- ⬤ Neutron
- ◐ Antineutron
- ○ Positron
- ○ Electron
- ∿➤ Photon

transparent. At this stage, which is known as the decoupling of matter and energy, the cosmic background radiation was released. As the radiation pressure was removed from the matter content of the Universe, the atoms began to succumb to the force of gravity, the atoms collected into vast clouds and the large-scale structure of the Universe began to evolve.

Between the release of the cosmic microwave background radiation and the present day, 15 billion years later, the Universe has expanded by a thousand times, and matter has collected and condensed to form galaxies, stars (including our Sun) and planets. As this has happened, the Universe has continued to cool.

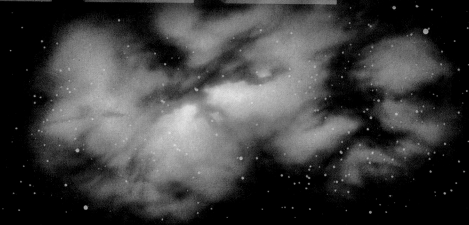

THE INFLATIONARY UNIVERSE

W HAT we see today as the observable Universe began as a region of space no larger than an atom. The Big Bang, widely believed to be the event that created the Universe, occurred between 10 and 15 billion years ago; what caused it is not yet understood. But astrophysicists have compiled an astonishingly detailed body of knowledge about what happened after the Big Bang, beginning only a microscopic fraction of a second afterwards, when the conventional laws of physics are thought to have been in place.

This tiny period of time following the creation of the Universe is known as the Planck era after the German physicist, Max Planck (1858–1947). It was during this era, which lasted 10^{-43} of a second, that a quantum theory of gravity is thought to be applicable.

In the very early Universe the four forces of nature – gravity, electro-magnetism, the strong nuclear force and the weak nuclear force – were combined into a single superforce. Matter and energy were not the apparently separate forces they are today. Even space was constantly being broken and folded up because of the incredibly small volume that the Universe occupied. As time passed, the Universe expanded. And as it expanded, the superforce separated into gravity and the Grand Unified Force.

The next crucial step in history took place when the Universe was 10^{-35} seconds old. By this time, it had expanded and cooled sufficiently for the Grand Unified Force to separate into the strong nuclear and the electroweak forces. Accompanying this separation was the sudden creation of quarks and leptons. This process was analogous

to the way in which water vapor in the atmosphere condenses into clouds when the temperature of the surrounding air falls sufficiently. Just as the change of water vapor into water releases heat energy, so the spontaneous creation of matter particles constituted a change in the Universe. This created enormous pressure, which drove the Universe to expand at a vastly accelerated rate – faster than the speed of light. This process, known as inflation, bloated the Universe by a factor of 10^{50}, despite the fact it all happened in just 10^{-32} seconds.

One of the most fundamental problems of understanding the Big Bang is that it is hard to visualize. This may be overcome by imagining the Universe as a flat piece of paper. In this model, our Universe is only two-dimensional, yet we know that a third dimension also exists. We can make use of this knowledge in the same way Einstein made use of a fourth dimension to explain gravity.

In the very early stages of the Big Bang, space was compressed so much that it had to curve through this extra dimension. In the two-dimensional model it would be the equivalent of screwing up the paper into a ball. This brings into contact regions that will have nothing to do with one another following inflation. Space in the early stages of the Big Bang was, essentially, folded in on itself. As the Universe expanded, the space unfolded. The only deformation which then took place in this extra dimension was the deformation caused by mass, creating gravity.

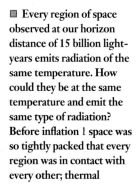

■ Every region of space observed at our horizon distance of 15 billion light-years emits radiation of the same temperature. How could they be at the same temperature and emit the same type of radiation? Before inflation 1 space was so tightly packed that every region was in contact with every other; thermal

equilibrium therefore existed. After the Universe had briefly "inflated" faster than the speed of light 2, objects such as quasars and galaxies formed; each one had its horizon, defined by the distance that light had traveled since the Big Bang. Thus A and B were now outside each other's horizon. In the modern

Universe 3, the same geometry applies, though the additional age of the Universe means that the horizons have expanded. In both stage 2 and 3, quasar A and B have no contact and cannot know of each other's existence, whereas we know that both exist because both quasars will stay within our horizon.

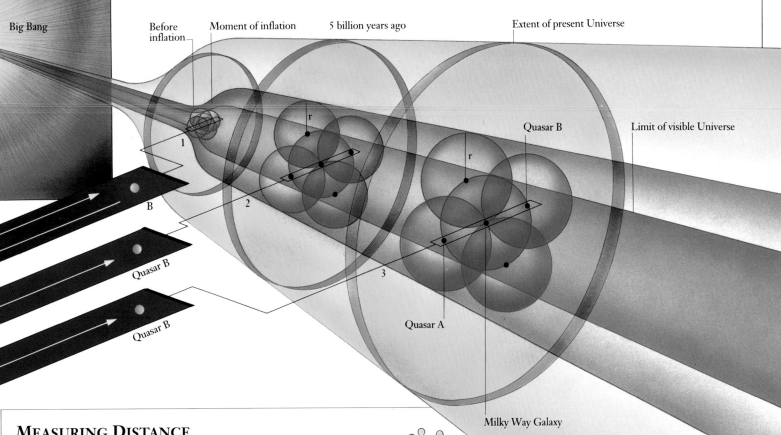

■ On Earth the horizon is the furthest point we can see because of the curvature of our world. In space, our horizon is the furthest point that we can see, due to the age of the Universe and the finite speed of light. If the Universe is 15 billion years old, our horizon is 15 billion light-years. Any two objects further than 15 billion light-years apart will not know of each other's existence, because the light emitted by each has not yet had time to reach the other. Before inflation, our horizon was expanding at the speed of light. At the point in which inflation took place, it only had a radius of 10^{-35} light-seconds. The space within it was bloated exponentially as the the Grand Unified Force broke up. Thus, the Universe became much larger than the observable portion. Regions that were once in contact with one another were carried apart by the expansion of space, which occurred at a velocity many times greater than that of light.

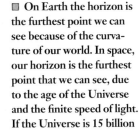

Big Bang

Before inflation

Moment of inflation

5 billion years ago

Extent of present Universe

Quasar B

Limit of visible Universe

B

Quasar B

Quasar B

Quasar A

Milky Way Galaxy

MEASURING DISTANCE

Astronomers use several units of length. Distances across the Solar System are measured in astronomical units (AU). This is the average distance between the Earth and the Sun – 1.496×10^8 km. One way of measuring the far greater distances between stars is in light-years (ly). A light-year is the distance traveled by light in one year – 9.46×10^{12} km or 63,240 AU.

Another unit, the parsec, is defined as the distance at which 1AU subtends an angle of 1 second of arc. (This is a very small angle; there are 60 seconds in one minute of arc and 60 minutes in 1°). One parsec equals 3.26 light-years.

The definition of the parsec is related to the method of measuring stellar distances known as parallax. As the Earth orbits the Sun, the position of nearby stars appears to move relative to more distant stars. Trigonometry is then used to calculate the distance.

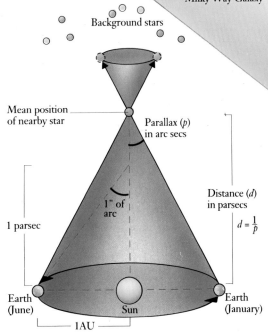

Background stars

Mean position of nearby star

Parallax (p) in arc secs

Distance (d) in parsecs

$d = \frac{1}{p}$

1" of arc

1 parsec

Earth (June)

Sun

Earth (January)

1AU

1 Visible horizons intermesh

2 Visible horizons separated: r = 10 billion light years

3 Visible horizons separated: r = 15 billion light years

The Infant Universe

B Y THE time the Universe was 10^{-12} seconds old, the electroweak force had separated to become the electromagnetic force and the weak nuclear force. Prior to this, all leptons – elementary particles such as electrons and neutrinos which are not made of quarks – had behaved in the same way. Now, however, with these two forces (which govern lepton reactions) separate from each other, electrons and neutrinos became distinct from one another. Electromagnetic interactions began to occur between all charged particles, and photons began to be created in abundance.

The constituents of the Universe at this stage were in a state of constant collision and interaction. Particles of matter collided with their antiparticles and were instantly annihilated into a pair of high-energy photons. These photons soon decayed back into particle-antiparticle pairs and the collision-annihilation process started again.

This back-and-forth conversion of matter and energy was possible because the Universe was so dense and hot: less than a millionth of a second after the Big Bang, the temperature was more than 10 million million K. In this environment,

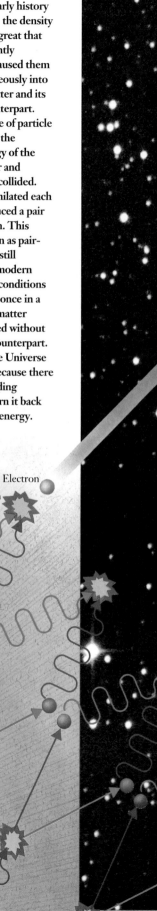

▽ In the very early history of the Universe, the density of space was so great that photons frequently collided. This caused them to turn spontaneously into a particle of matter and its antimatter counterpart. The precise type of particle depended upon the combined energy of the photons. Matter and antimatter also collided. They then annihilated each other and produced a pair of photons again. This process is known as pair-production and still happens in the modern Universe when conditions are right. Every once in a billion times, a matter particle is created without an antimatter counterpart. This "seeds" the Universe with particles because there is no corresponding antimatter to turn it back into photons of energy.

Electron

Photon

Electron

Positron

Big Bang

quarks could exist as separate particles because any bonds they did make with other quarks were instantly broken by collisions.

When the Universe reached one microsecond in age, conditions changed. By now it had expanded and cooled sufficiently so that the spontaneous creation of matter was no longer possible on the same vast scale it once had been. By this time, when the particles and antiparticles collided, the resulting photons did not turn back into matter.

As the Universe cooled, the strong nuclear force pulled the quarks together to assemble as protons and neutrons. Most of these were annihilated in turn as they collided with their antimatter counterparts. However, because of a small but measurable tendency in the Universe to create more matter than antimatter, a residue of elementary particles remained. For every one billion particle-antiparticle pairs, one particle was created without an antimatter counterpart. This residue of matter particles makes up every atomic nucleus found in the Universe today.

The neutrinos and antineutrinos had been in a constant state of collision with the other constituents of the Universe up to this point. As the Universe reached one second in age, they all but ceased to react with the other particles. This process, called neutrino decoupling, is potentially one of the earliest detectable events after the Big Bang: if there were sufficiently powerful neutrino detectors, it could be detected as a background flux of neutrinos, allowing astronomers to study the Universe as it existed at one second in age.

The only earlier event that is potentially detectable was graviton decoupling, which is believed to have occurred 10^{-12} seconds after the Big Bang. However, graviton decoupling is even more speculative than neutrino decoupling: unlike neutrinos, gravitons are still not proven to exist at all.

◁ **All matter in the Universe (including these stars in open star cluster NGC 3293, shown here) is composed of particles of matter that were created without a corresponding particle of antimatter. Photons dominate the Universe over particles of matter by a ratio of a billion to one. The first stars in the Universe were composed of only hydrogen and helium. The heavier elements had yet to be synthesized** because this process occurs only in the centers of massive stars. Only when the first generation of stars reached the ends of their lives could they seed space with elements heavier than helium. It is thought that galaxies began to form about 1 billion years after the Big Bang. It is impossible to know exactly, however, because these were much farther away than even the most distant objects we can detect.

BEGINNINGS OF STRUCTURE

As the Universe expanded, fractions of seconds after the Big Bang, its temperature continued to fall. When it was about fifteen seconds old, the temperature had fallen sufficiently to prevent electron-positron pairs forming spontaneously. In the same way that the neutrons and protons, and their antimatter counterparts, had annihilated one another to leave a small residue of matter, so the electrons and positrons now did the same. Again, the small bias towards matter ensured that for every one billion electron-positron annihilations, one electron survived. This now meant that, for every particle of matter, several billion photons also existed.

Although the Universe was still dominated by photons and neutrinos, now it had also been seeded with the building blocks of atoms (protons, neutrons and electrons). The Universe's entire quota of elementary particles was now in place and they existed in a state of constant collision.

As the Universe became one minute old, the conditions became right for neutrons and protons to begin to assemble into atomic nuclei (nucleosynthesis). This became possible because the collisions that now occurred, especially those between the baryons (neutrons and protons), were much less violent as a result of the Universe cooling and the particles no longer moving at such high speed, permitting the strong nuclear force to take effect on particles as they came into contact. Collisions between protons and neutrons built atomic nuclei through the process of nuclear fusion.

After approximately four minutes of nucleosynthesis, the Universe had expanded sufficiently, and the temperature had dropped correspondingly, to stop the process. By now the "primordial elements" (those created in the infancy of the Universe, as opposed to those created later in the cores of stars) were made. The Universe now contained atomic nuclei of hydrogen (a single proton) and its isotopes deuterium (one proton with one neutron) and tritium (one proton and two neutrons), helium (two protons and two neutrons) and its isotope helium-3 (two protons with a single neutron).

Individual neutrons unattached to protons ceased to exist, because neutrons require the presence of other baryons if they are to remain stable. Outside the confines of an atomic nucleus, therefore, neutrons decay into a hydrogen nucleus (a single proton) and an electron. The difference in mass between the neutron and the combined masses of the electron and proton would have been converted into energy and carried away by a neutrino.

The Universe was still too energetic for the electromagnetic force to be able to bind electrons to atomic nuclei. Any electrons that were captured by a nucleus were swiftly supplied with enough energy by collisions with photons to escape from them again. The Universe remained in this permanent state of ionization for several hundred thousand years.

At some age between 300,000 and 500,000 years, one of the most profound changes occurred in the Universe: the so-called decoupling of matter and energy. As the expansion of the Universe lowered its temperature, it became ever more difficult for photons to knock electrons away from nuclei. As the electrons were attracted to atomic nuclei, photons became able to travel large distances without colliding with other particles. In a sense, therefore, the Universe became transparent to the photons within it.

The radiation released during this event is detectable today as cosmic microwave background radiation, redshifted enormously by the expansion of the Universe. It is remarkably consistent across the sky, representing a temperature of 3 K, and is the direct afterglow of the huge temperatures that occurred 15 billion years ago in the Big Bang.

The decoupling of matter and energy, which gave rise to this radiation, is the earliest observable event in the Universe, and the discovery of background radiation in 1965 provided the first decisive evidence for Big Bang theory. In the late 1980s, observation of the minute variations in this radiation – less than one part in 10,000 – using the COBE satellite provided even more important evidence that the Universe was not uniform at this point, and consisted of hotter but thinner regions and cooler but denser regions of space.

Once the collisions between matter and radiation stopped, the force of gravity – so much weaker than the other forces – could pull the atoms together. This meant that the large-scale structure of the Universe

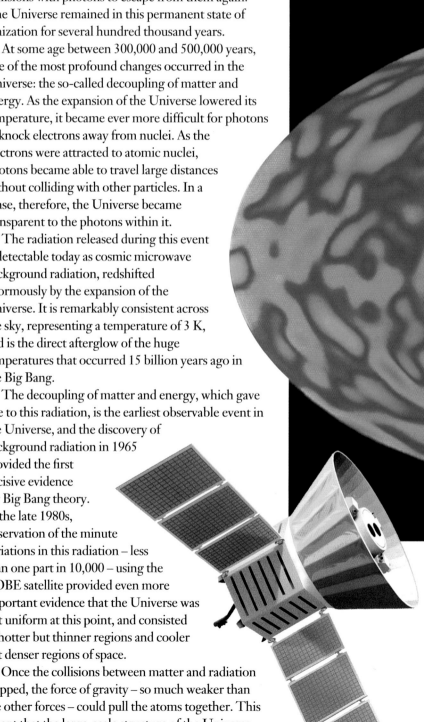

▼ The COBE (COsmic microwave Background Explorer) satellite was launched by NASA in the late 1980s. The discovery of the cosmic background radiation provides historical evidence that supports the Big Bang theory.

▷ United States astrophysicists Arno Penzias and Robert Wilson discovered the cosmic microwave background radiation in 1965 using this radio horn at Bell Laboratories, New Jersey. They subsequently won Nobel Prizes.

▽ An infrared image of an insulated house shows the principles that are used on the COBE map. Where the insulation is most dense, the image is bluest (coolest) because it is most difficult for heat to escape.

▲ An all-sky map of the microwave background radiation was taken by the COBE satellite. The areas are color-coded to indicate temperature. Blue regions are cooler than pink ones. The blue areas appear cooler because they contain more material and radiation cannot easily escape.

began to evolve. Although astronomers have not yet fully resolved the details of this process, it is likely that clouds of atoms came together to form the various galaxies that dominate our view of the cosmos, and eventually areas within the clouds collapsed further to form stars with nuclear fusion at their cores. However, there is no direct evidence that this is true, because the oldest known quasars date from about 2 billion years after the Big Bang; astronomers have not yet been able to detect any objects older than these.

3

GALAXIES
& Quasars

THE GALAXIES began to emerge, some 500,000 years after the Big Bang, from huge clouds of matter collapsing under the force of their own gravity. As they took shape, the material that they had accumulated began to collapse as well. As this fragmentation continued, the smaller lumps of matter formed stars. Thus galaxies formed at the same time as the stars they contained.

Galaxies are huge assemblages of stars and other less luminous matter. Some galaxies contain nothing but old stars; others have regions in which new ones are continually created. Galaxies can contain many millions of stars and are found in many shapes.

The Sun belongs to the Milky Way galaxy, which has stars distributed in a flat spiral disk. Every star visible with the unaided eye also belongs to our Galaxy. The misty band of light that stretches across the night sky is the combined light from the other, more distant stars in the disk of the Milky Way.

Only three galaxies are visible with the naked eye. Two, the Large and Small Magellanic Clouds, are satellite galaxies to our own. The third appears like a dim star but is a nearby spiral galaxy known as the Andromeda galaxy. Most galaxies astronomers observe through telescopes are at extreme distances; they look like misty swirls and patches of light and their individual stars cannot be made out.

All the stars within the Universe are contained within galaxies. A galaxy is a collection of millions of stars, and there are countless millions of galaxies. This galaxy, known as M74, is a typical spiral galaxy, 30 million light-years away from Earth and measuring 80,000 light-years in diameter – about the same size as the Milky Way, the spiral galaxy of which the Earth's Solar System is part. Other galaxies may be a hundred times smaller; still others, more than five times larger.

CLASSIFICATION OF GALAXIES

Galaxies are found in a great variety of shapes and sizes, but they can be classified into two main types just by looking at them. Nearly all galaxies are either elliptical or spiral in appearance.

Classification is normally made according to shape, following a scheme known as the "tuning fork" diagram, first devised by the American astronomer Edwin Hubble in the 1920s. Elliptical galaxies are huge collections of stars that range in shape from perfect spheres to flattened ellipses which resemble cigars. The largest galaxies in the known Universe are enormous elliptical systems. They exist at the centers of dense clusters of galaxies, and are estimated to contain up to one hundred billion stars.

It seems likely that these galaxies grew so large by absorbing smaller ones which strayed too close and became caught in their vast gravitational fields. On the other hand, dwarf elliptical galaxies are some of the smallest star systems known, with only about a million stars. They are thought to be abundant, but are difficult to detect because of their small size. All stars contained in elliptical galaxies are old, and there is no star formation currently taking place within them.

Spiral galaxies are beautiful objects, resembling pinwheels, which show definite signs of recent and continuing star formation. They contain a central bulge of older stars known as the nucleus, surrounded by a disk of material in which new stars are constantly forming. Where stars have formed in the disk, they shine with brilliant intensity and trace out spiral patterns around the nucleus. These spiral "arms" gradually rotate around the galaxy, following the compressed regions of disk material within which new stars are forming.

Spiral galaxies come in a variety of types, which are normally classified according to how tightly wound the spiral arms are and how large the nucleus is. Approximately half of all spiral galaxies identified so far have an additional distinguishing feature. This is a straight barlike structure of stars that emanates from the galactic nucleus and stretches into the disk. The conventional spiral arms then twist around from the bar ends. These galaxies are called barred-spiral galaxies. Like spiral galaxies, they can be subdivided into different types according to how tightly wound their arms are and how large the nucleus is. The origin of the bars appears to relate to the gravitational interactions of the stars in a rotating spiral.

Lenticular galaxies form an intermediate class of galaxy, between ellipticals and spirals. They have nuclear bulges and a thin disk of stars, but they lack spiral arms. Sometimes lenticular galaxies also have a barlike structure.

Galaxies with no obvious structure or nucleus are called irregular galaxies. Type I irregulars are galaxies which show evidence of spiral arms that have been disturbed in some way. A Type II irregular is just a confused jumble of stars. There is evidence that very small galaxies of this type, known as dwarf irregular galaxies, can be formed from the matter flung into intergalactic space during collisions between larger galaxies. Like spirals, irregulars are still undergoing the process of star formation.

▷ **Galaxies are the largest single objects in the Universe, measuring an average of 100,000 light-years across. M83 is a spiral galaxy in the constellation of Hydra. It has two** **obvious spiral arms and a fainter third one. M83 is located about 27 million light-years away from our own galaxy, the Milky Way, and has a diameter of 30,000 light-years.**

△ **The Hubble "tuning fork" diagram shows several different types of galaxy. There are seven classes of elliptical galaxy 1-3, depending on how flat they are. Spiral galaxies 4-6** **and barred-spiral galaxies 7-9 are usually shown on the right-hand side of the diagram. Spiral galaxies are subclassified into three types, depending on the size of the nucleus and how** **tightly wound the arms are. Lenticular galaxies are often placed between the spiral and elliptical galaxies. Galaxies that do not fit any other classification are called irregulars.**

4

5

6

7

8

9

◁ Galaxies were once thought by astronomers to form as elliptical shapes and gradually become flatter as they rotated. It was believed that the galaxies then developed spiral arms and became spiral or barred-spiral galaxies. However, it is now known that this does not happen – in other words, the different types of galaxies represented on the Hubble tuning fork diagram are not an evolutionary sequence. The Hubble classification of a galaxy never changes unless the galaxy undergoes some type of catastrophic event such as a collision with another galaxy. One theory states that elliptical galaxies can be formed when spiral galaxies collide and merge.

STRUCTURE OF GALAXIES

THE visible regions of a spiral galaxy were once thought to represent the system in its entirety. Astronomers now believe that the matter which has formed stars is no more than a tiny fraction of the total material contained within a galaxy. This other mass is contained in the form of faint objects, which are too dim to see from the distances we view galaxies, or other forms of matter that we cannot directly detect.

Among the matter that is too dim to see from Earth, the disk of a spiral galaxy contains vast lanes of dust and gas which are not illuminated. Sometimes dust lanes become visible because they block out the light from spiral arms and allow us to see them in silhouette form. The galactic disk also contains many fainter, older stars which cannot be seen because they are outshone by the young, bright stars in the spiral arms. The rotation of stars around spiral galaxies has provided important clues that the galaxies contain much more matter than can be seen. Studying the way in which spiral arms rotate has led astronomers to believe that huge, hidden spherical halos of matter exist around spiral galaxies.

From the visible evidence it would appear that the bulk of the mass of a galaxy, like the mass of the Solar System, is concentrated in its core. This would imply that, as the galaxy rotates, the stars that are furthest from the core would move more slowly than those that are closer to the core. Observation does not bear this out, however. Instead, it is more likely that the bulk of the mass of a galaxy exists beyond its visible limits, contained in a vast spherical halo of matter.

Matter in the halo is thought to be contained in a number of different objects such as dim stars that have escaped the disk of the galaxy; failed stars, known as brown dwarfs; and the remains of stars that have collapsed and died, forming objects including neutron stars and black holes. Gas clouds are probably present in the halo as well. Together with the dimmer objects, the halo also contains luminous ones known as globular clusters.

Globular clusters can be thought of as small cousins to elliptical galaxies. They are spherical conglomerations of stars held together by their mutual force of gravity. No star formation is taking place within the globular clusters. They orbit the nucleus of their parent galaxies, and define a spherical region thought to indicate the limits of the galactic halo.

Globular clusters contain stars that are very old – most are thought to have formed about 10 billion years ago. Some globular clusters are much older, however, with estimated ages as old as the Universe itself. The largest globular clusters contain a few million stars. Spiral galaxies typically have around 150 globular clusters, whereas elliptical galaxies can have up to one thousand. It is thought that when gas clouds collapsed to form galaxies, isolated regions collapsed separately and formed globular clusters.

Many astronomers believe that beyond the galactic halo there exists an even larger spherical region, known as the corona. The corona may be as much as four times as large as the diameter of the galactic halo. It may contain exotic particles, known as dark matter, which behave very differently from the five stable elementary particles. Although these exotic particles are, at present, undetectable because of the limitations of even the most state-of-the-art technology now available, their presence can be nevertheless inferred because of their gravitational effect on the luminous matter in the galaxy. It has been suggested by some astronomers that the corona could contain up to 90 percent of the total matter of the galaxy.

▷ In a turning wheel, the whole thing turns as one. The outer edge travels fastest, because it has farthest to go. In the Solar System, with the masws at the center, the outer planets travel more slowly. Stars in the Milky Way galaxy follow the middle line, with most stars in the disk possessing the same orbital velocity. This shows a substantial pull of gravity from matter in the halo.

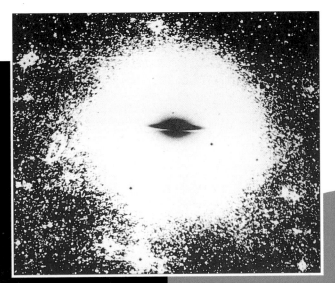

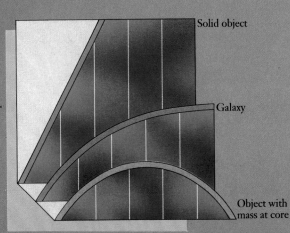

The visible regions of a spiral galaxy are part of a much larger structure. Here a typical spiral galaxy is seen edge on; the disk structure is surrounded by the halo, in which globular clusters are conspicuous. In addition to these, the halo is thought to contain dim stars, dead stars such as white dwarfs and neutron stars, and even black holes.

Beyond the halo of a spiral galaxy, some astronomers believe that an even larger spherical region of material exists. This volume is known as the galactic corona and contains, according to current theory, a large amount of dark matter. As yet, nobody has detected this material but its existence is inferred by the motion of galaxies within clusters of galaxies. Coronal dark matter may explain oddities in the rotation of galaxies.

Globular clusters helped an American astronomer, Harlow Shapley, to make the first accurate measurement of the Milky Way Galaxy, in 1920. It is difficult to observe an entire galaxy; interstellar dust in the galactic plane restricts the view. Globular clusters (on the ends of the yellow lines) are located above and below the plane, where there is less dust. Shapley assumed that the center of the system of clusters coincided with the center of the galaxy, and used the distances to the clusters to estimate the size of the Milky Way.

▤ The Sombrero galaxy (M104), in the constellation of Virgo, is an edge-on spiral galaxy LEFT. The dark stripe across the middle is due to dust. Sophisticated computer-image processing has made the faint halo visible ABOVE. A "negative" of the galaxy is then superimposed to reveal its location.

◁ M13 is a globular cluster associated with the Milky Way. Such clusters exist in the halos that surround galaxies and orbit the nuclei of their parent galaxy. In spiral galaxies, these orbits cause the clusters to pass through the disk regions. The density of stars is so low, however, that the globular cluster emerges on the other side intact.

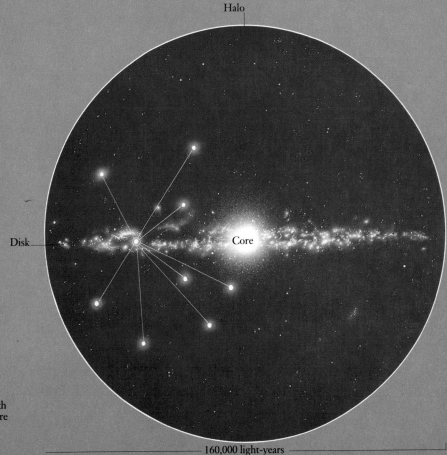

Corona

Halo

Disk

Core

Rotation speed

Solid object

Galaxy

Object with mass at core

Distance from center

160,000 light-years

THE MILKY WAY

TRADITIONALLY, when people talk of the Milky Way, they are describing the misty band of light which stretches across the night sky. The Italian astronomer Galileo (1564–1642) was the first person to look at the Milky Way with a telescope. He saw that it was composed of countless faint stars. During the next three centuries astronomers came to realize that this faint band of light is our view of our own galaxy. The reason it looks so different from the other galaxies we can see is that we are viewing it from within.

The Milky Way is a spiral galaxy and is therefore relatively flat and disklike. If we look along the plane of the disk, we see many more stars than we do by looking to each side. The Sun, however, is not at the center of the Milky Way but is located in one of the spiral arms. The center of the galaxy lies in the direction of the constellation known as Sagittarius, and the spiral arms are named after the constellations (patterns of stars) through which they pass.

Although the galaxy formed between 10 and 15 billion years ago, the Sun was only formed in a spiral arm some 4.5 billion years ago, and has been in orbit around the center of the Milky Way ever since. It has completed roughly 21 orbits and is currently situated on the trailing edge of the Orion arm. As the name implies, this is the arm that contains most of the stars in the constellation of Orion. Some recent work on mapping the galaxy has suggested that Orion may not, in fact, be a complete spiral arm but simply an off-shoot that connects the Sagittarius arm and the Perseus arm. If this is the case, our location would be more correctly described as being within the Orion bridge or spur. The Sagittarius arm lies between us and the galactic center, whilst the Perseus arm curls around outside the Sun.

The galactic center itself is a place of considerably mystery. It is shrouded in dust and gas clouds which block a clear view of what it contains. Visible light cannot penetrate these clouds, and astronomers rely on observations at other wavelengths of electromagnetic radiation which can travel through the dust clouds. One of the brightest sources of radio emission in the sky comes from an object known as Sagittarius A*. It lies at the galactic center and many astronomers believe that it is an exotic object known as a black hole.

▷ **The center of the Milky Way galaxy is located in the direction of the constellation Sagittarius (which is shown here). The high density of stars visible indicates how tightly they are packed. Our own direct view of the central region is** blocked by the vast amounts of dust in the galactic disk between the Earth and the galaxy's center. However, at wavelengths other than those of visible light, the center of the Milky Way can be revealed.

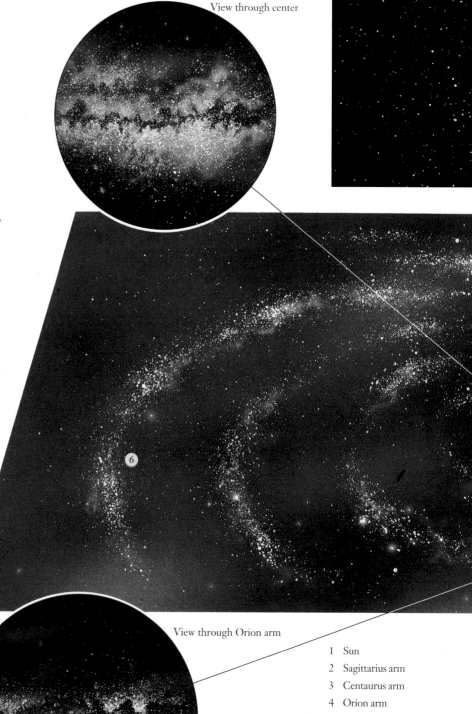

View through center

View through Orion arm

1 Sun
2 Sagittarius arm
3 Centaurus arm
4 Orion arm
5 Perseus arm
6 Cygnus arm
7 Center of Galaxy

Without doubt, the Milky Way is a spiral galaxy. Which type of spiral galaxy it is, however, remains contentious. For many years it was believed to be a standard spiral galaxy. But there may be a small bar joining the nucleus to the spiral arms, so the Milky Way is a barred spiral galaxy. Another interesting feature of our galaxy's shape is that the disk of stars is not flat but warped.

Like many large galaxies, the Milky Way has a number of smaller galaxies in orbit around it. The Magellanic Clouds are two irregular satellite galaxies, and there are a number of even smaller, dwarf galaxies caught in its gravitational influence. Beyond its overwhelming influence, the galaxy is gravitationally bound to others in an association of galaxies known as the Local Group. This contains 21 known members, of which three are spiral galaxies (the Milky Way, Andromeda and M33). The rest of the galaxies in the Local Group are ellipticals – including the giant elliptical Maffei I – and dwarfs.

▽ **Long-exposure photographs such as this show how the density of stars increases and the thin disk of the Milky Way widens into the elliptical bulge of stars known as the galactic nucleus.**

The picture also shows several of the dust lanes present in the galactic disk. Careful analysis of the photograph shows that the density of globular clusters is greatest in the region surrounding the nucleus.

▲ **In this stylized view of the Milky Way, a number of key features have been represented to show why different views from the Earth make the Milky Way appear differently to us. No matter which way we look,** the view is always a super-imposition of spiral arms. The Milky Way appears densest when we look towards the galactic center. Other views pass through different volumes of stars – some more, some less.

CLUSTERS AND VOIDS

Almost every galaxy is associated with other galaxies through the force of gravity. These associations are known as groups or clusters, depending upon how many galaxies are involved. Our own galaxy, the Milky Way, is a member of the Local Group, which contains about 20 galaxies of various sizes. Associations with more than a few dozen galaxies are called clusters. They are found in a variety of shapes and sizes. Some are spherical; others – such as the Milky Way – are irregular and sprawl across space. The different types contain different types of galaxy. By studying the types of galaxy contained, astronomers can understand how the shapes of galaxies evolve – especially the way spiral galaxies become elliptical galaxies.

In spherical clusters, most of the galaxies are elliptical. These clusters resemble globular clusters but on a vastly larger scale. Instead of being composed of individual stars, they are made of individual galaxies. These cluster around a central concentration of galaxies and follow well-defined elliptical orbits, which regularly take them into these dense regions. Once there, spiral galaxies collide with each other and become elliptical galaxies. In some cases the central object of a cluster is a giant elliptical galaxy. These are known as cluster-dominating, or cD, galaxies. They are thought to have been produced by successive mergers of several smaller galaxies. Irregular clusters contain mostly spiral galaxies and have no well-defined shape or center of gravity. Their constituent galaxies rarely come into contact with one another.

Clusters of galaxies can also be bound to other clusters by gravity. During the 1980s astronomers mapped the positions and red shift (and therefore the distance away) of thousands of galaxies, which suggested that these are not spread uniformly across space, but build up into chains called superclusters which sweep across the Universe. The Local Group is thought to be part of a local supercluster, called Virgo, which has a diameter of more than 100 million light years. Superclusters seem to be concentrated around huge spherical voids. This may relate to the "lumpiness" in primordial matter detected by surveys of the cosmic background radiation. The largest supercluster is a thin sheet called the Great Wall, which seems to cover more than 250 million by 750 million light years.

Gravity in clusters often overcomes the expansion of the Universe. The galaxies then move according to their gravitational attraction to each other. But superclusters are so large, the space between them must be expanding, as shown by the Hubble flow. This cannot be simple expansive motion, because of the force of gravity. Instead of growing uniformly, the superclusters oscillate gently as they expand with the Universe.

▷ **The Local Group is part of the Virgo (or Local) supercluster. About 20 percent of the supercluster's member galaxies are found within the Virgo cluster. This cluster, about 50 million light-years away, consists of a thousand galaxies contained within a region measuring about 7 million light-years across.**

◁ **In the central part of the Virgo cluster are some of the nearest neighbors to the Local Group. The giant elliptical galaxies shown here measure 2 million light-years in diameter – each one is almost the same size as the entire Local Group of galaxies.**

▷ **The Milky Way is only one galaxy out of the 20 or so that make up the Local Group of galaxies. The number 20 is a conservative estimate, because it is almost certain that many galaxies with a fainter appearance remain to be discovered.**

Virgo Supercluster

1 Virgo III cloud
2 Virgo II cloud
3 Crater cloud
4 Virgo I cloud
5 Leo II cloud
6 Canes Venatici cloud
7 Canes Venatici spur

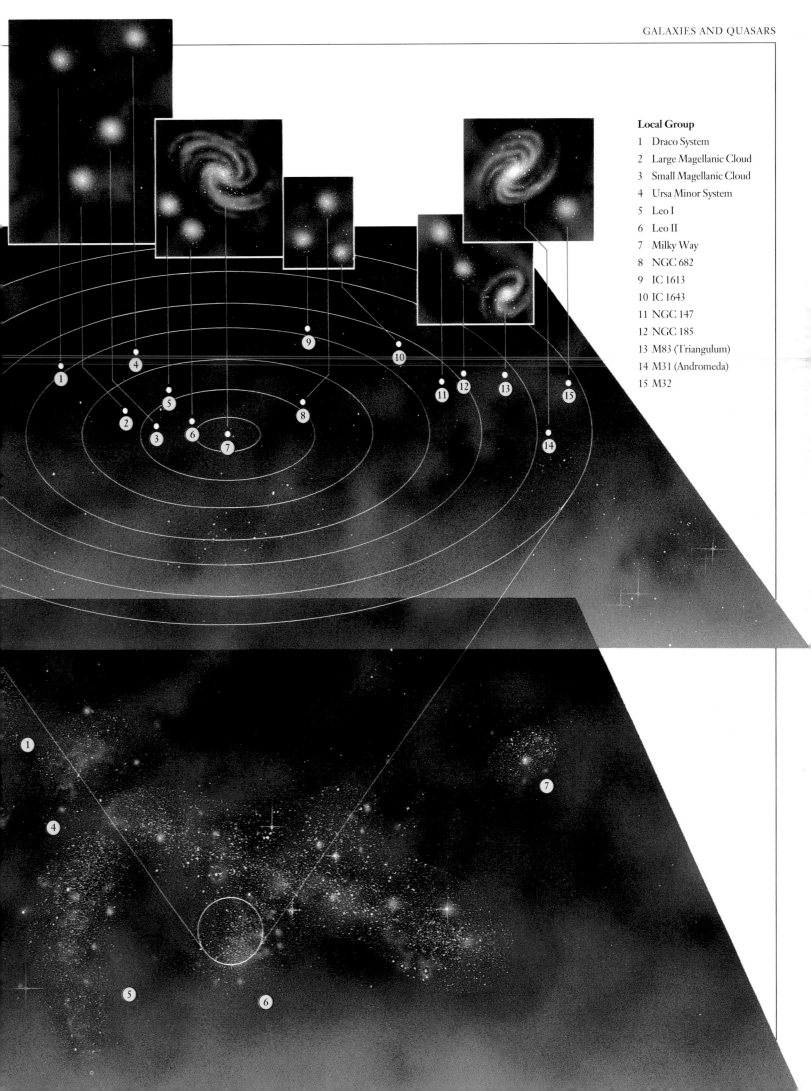

Local Group

1 Draco System
2 Large Magellanic Cloud
3 Small Magellanic Cloud
4 Ursa Minor System
5 Leo I
6 Leo II
7 Milky Way
8 NGC 682
9 IC 1613
10 IC 1643
11 NGC 147
12 NGC 185
13 M83 (Triangulum)
14 M31 (Andromeda)
15 M32

ACTIVE GALAXIES

ALTHOUGH strange and energetic phenomena have been detected at the center of the Milky Way, they are in no way comparable to those observed in the so-called active galaxies. Ten percent of galaxies are termed active. The nucleus of an active galaxy is often so bright that it outshines the starlight from the rest of the galaxy. There are many types of active galaxy, each with its own peculiar properties.

The first type to be found became known as a Seyfert galaxy, after its discoverer, Carl Seyfert. Seyfert galaxies are spiral or barred-spiral galaxies that have very bright nuclei. When analyzed spectrographically, Seyferts show strong spectral emission lines produced by clouds of hot gas. Seyferts are strong sources of infrared radiation, although not all of them emit radio waves. Type I Seyferts have spectral emission lines that indicate they are being produced by clouds of hydrogen swirling at very high speeds around the center of the galaxy. Type II Seyferts, although displaying hydrogen lines, do not seem to have fast-moving gas clouds.

Quasi-stellar objects (QSOs) are thought to be very similar to Seyfert galaxies except that the activity in their nuclei is much greater. They appear as star-like points of light in the sky (hence the term quasi-stellar object), but are obviously not stars when studied spectroscopically. They exist at extreme distance, as indicated by the redshift of their spectral lines, and are among the most distant objects known in the Universe. Like Seyfert galaxies, they can be "radio loud" (in which case they are termed "quasar" which stands for quasi-stellar radio source) or "radio quiet" (the traditional QSO). The luminosity of a quasar can be a thousand times the luminosity of a normal galaxy. The surrounding galactic structure of the quasar has yet to be observed because it is so much dimmer than the active nuclear region; this is why the various types of active galaxy are commonly called active galactic nuclei (AGNs).

Another form of active galaxy is the radio galaxy. As the name implies, such galaxies emit most strongly in the radio region of the electromagnetic spectrum. Instead of a point source, the emission comes from vast radio lobes, located on each side of the parent galaxy. A typical spiral galaxy has a diameter of approximately 100,000 light-years and yet, from lobe to lobe, a radio source can span tens of millions of light-years.

The final type of active galaxy is known as a blazar. Known also as BL Lacertae objects, blazars are similar to quasars in most respects, except that they display no spectral lines.

The fact that most active galaxies are at extreme distances indicates that they are young objects in terms of the history of the Universe, because their light has taken millions of years to reach us. This, in turn, leads astronomers to believe that perhaps all galaxies go through this active phase.

KEYWORDS

ACTIVE GALACTIC NUCLEUS (AGN)

ACTIVE GALAXY

BLAZAR

ELECTROMAGNETIC SPECTRUM

GALAXY

LUMINOSITY

MILKY WAY

QUASAR

RADIO GALAXY

REDSHIFT

SEYFERT GALAXY

SPECTRAL EMISSION LINES

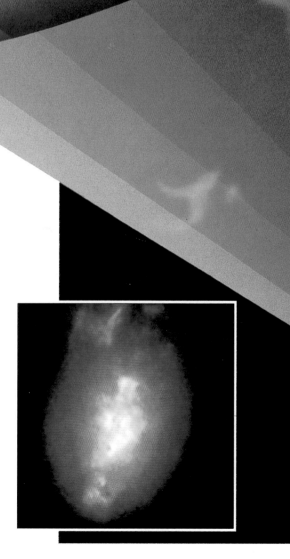

■ **NGC 1068, also known as M77, is an active galaxy in the constellation of Cetus. Even simple observations show that this celestial object is a Seyfert galaxy. More detailed study led to its classification of that of a Type II. However, recent observations of its nuclear region in scattered light RIGHT suggested the presence of fast-moving hydrogen clouds, characteristic of Type I galaxies. This led to the theory that all Seyfert galaxies are the same, but in some the clouds are obscured by a thick ring of dust surrounding the center of the galaxy. The image INSET taken from the Hubble Space Telescope provided new detail of this central region.**

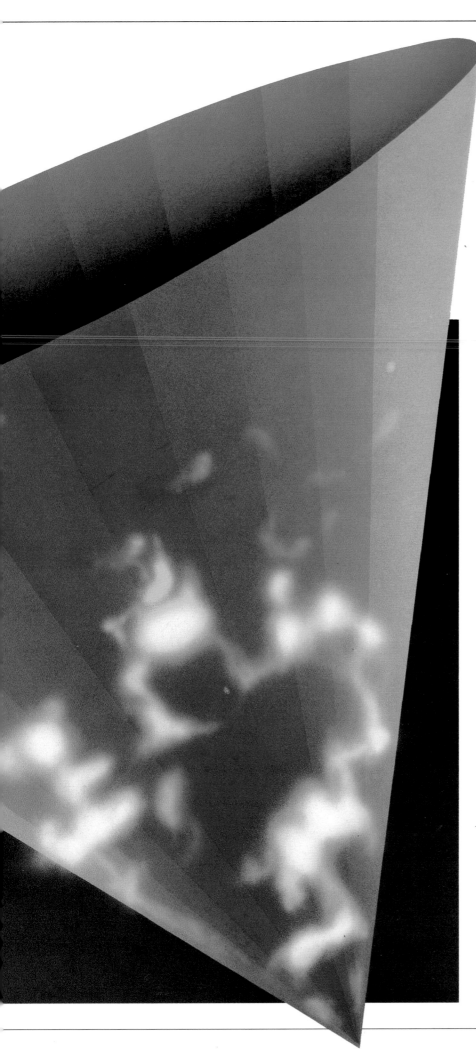

△ **NGC 4151** is another Seyfert galaxy, as can be seen from its very bright nuclear region. Seyferts are, without exception, spirals with extremely bright nuclei; their gas clouds at their centers may move at 5000 kilometers a second.

▽ **Quasars**, such as 3C 273, shown here, are much farther away than Seyferts, and also are much brighter. They are the farthest visible objects in the known Universe. This image shows X rays streaming from the galactic center.

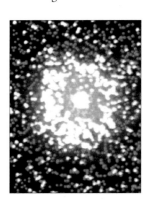

▽ **Radio galaxies** exist in the same region of space as Seyferts. The radio lobes of Centaurus A, seen here, extend for almost 2.5 million light years to either side of the galaxy itself, which is visible as the pink and red region in the center of the image.

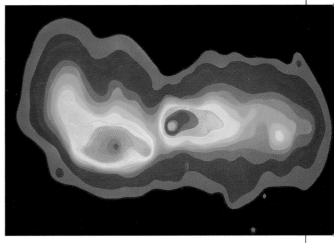

THE ENERGY MACHINE

THE various types of active galaxies – Seyferts, quasars, radio galaxies and blazars – all seem very different from each other. However, many astronomers now believe that they are fundamentally the same type of object. The reason they look so different is because, from Earth, we view them at different angles.

Active galaxies need some form of central "engine" to generate the enormous amounts of power they radiate. Although there are many processes that can generate vast amounts of energy, an object falling into a gravitational well is the most efficient. This has led most astronomers to believe that, in the heart of active galaxies, energy is released because matter is being sucked down into a supermassive black hole. Black holes are objects in space that are so dense that nothing can escape from them – not even light itself. A black hole has such a powerful gravitational field that any stars, gas clouds or other matter that strays too close is swallowed up and never seen again.

The material falling into the black hole does not travel straight downward. The rotation of the galaxy causes the material to be strung out into a disk, known as an accretion disk, from which the material or object can complete its journey into the black hole. The matter in the accretion disk rotates very rapidly, causing it to heat up and emit X rays and other forms of electromagnetic radiation. Because the accretion disk is so thick, the radiation cannot readily escape through it. Instead, it is beamed along the axis of the accretion disk, where it encounters the least resistance to its passage. Subatomic particles are also accelerated along the axis, forming jets. These collide with atoms in the intergalactic medium and shock them into emitting radiation at radio wavelengths. It is these radio-emitting lobes that are detected in radio galaxies.

Surrounding the accretion disk is a doughnut shape of dust and gas known as a torus. The torus is heated by the short-wave emission coming from the accretion disk. The material in the torus then re-emits this radiation at longer wavelengths. Gas clouds, which swirl around the central engine, are also heated by emission from the accretion disk and emit radiation detectable as spectral lines.

The difference between a Seyfert galaxy and a quasar is merely the intensity of radiation produced in its nucleus. The difference between these and other types of active galaxy is to do with viewing angle. In an active galactic nucleus viewed along the accretion disk, the bright central engine is obscured by the surrounding torus. Only the radio lobes are visible, and we "see" a radio galaxy. If, however, the object is observed along the disk axis, it is possible to look straight down the jet, where the intensity of radiation is dazzling. As hot gas is accelerated along the jet, it causes the brightness to vary, resulting in the characteristic form that appears to viewers as a blazar. At viewing angles between those of a blazar and a radio galaxy, emission from the central engine and possibly the jets is observed. Also, emission from the hot gas clouds is visible. This object is a quasar.

This unified theory of active galaxies is not yet proven, but it does present a very attractive theory.

▷ According to the grand unified theory of active galaxies, the orientation of the active galactic nucleus determines the way it looks from the Earth. If the disk is face-on 1, we look directly down the jet, which is so bright that we cannot see the disk itself. At others angle 2 the jet does not point directly at us and we should be able to see the disk that surrounds the black hole. When the disk is edge-on to us 3, we cannot see the activity taking place in the vicinity of the black hole.

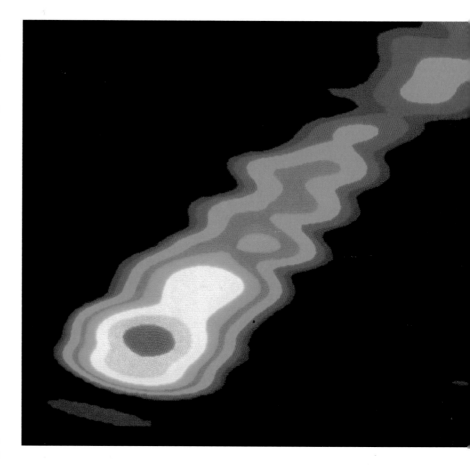

▲ A radio image of quasar 3C273 shows a jet coming from the active galactic nucleus (red, lower left). The jet is a beam of fast-moving subatomic particles. Although 3C273 is the nearest known quasar, it is still 2.1 billion light-years from Earth.

▷ The diagram shows what most astronomers believe is happening in the center of an active galaxy. A large region of dust is distributed into a disk, known as a torus. At the center of the torus lies a black hole. Clouds of hydrogen gas orbit the black hole. By a mechanism which is still not understood, a bipolar jet of subatomic particles streams out of the active galactic nucleus at right angles to the plane of the dust torus. The diagram can be compared with the photograph of this region in NGC4261 TOP RIGHT .

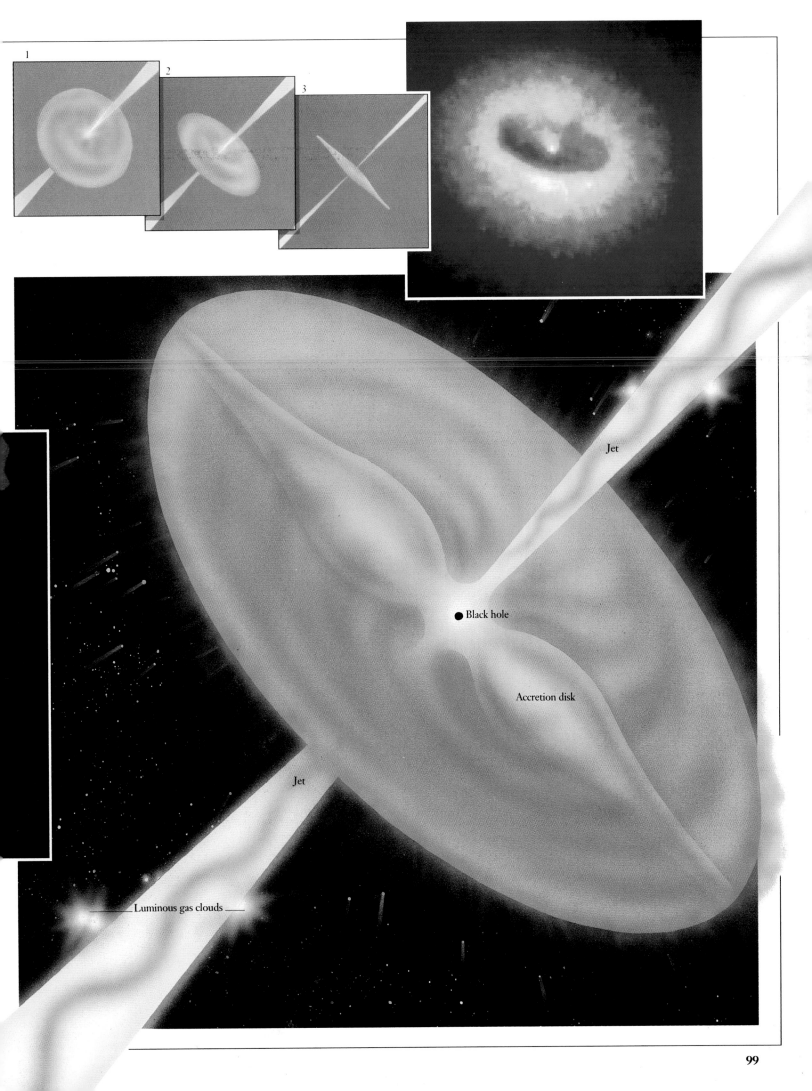

Jet

Black hole

Accretion disk

Jet

Luminous gas clouds

INTERACTING GALAXIES

GALAXIES in a cluster are constantly in motion (because of the mutual gravitational force between them and their neighbors), and so from time to time, perhaps once every several million years, they pass so close to each other that dramatic interactions can occur. If the two galaxies are of similar masses, the results of the interaction are very different from those that occur when one galaxy is much larger than the other. The proximity of the galaxies also affects the final outcome. Some galaxies pass each other and make their presence felt from a distance, whereas others plunge together and merge.

If two spiral galaxies of similar mass fall toward each other, as they draw ever closer they begin to disturb each other's stellar content. They pull each other's stars from their orbits, and slowly the galaxies lose their spiral patterns. Some of these stars are pulled away from the galaxies and strung out in long "tails" through intergalactic space. Other stars are decelerated and begin to fall toward the center of mass of the two galaxies. If the galaxies pass sufficiently close, they merge and become one. When galaxies collide in this way, the stars they contain do not actually touch each other: the spaces between the stars are so large that the chances of collision, even in a galactic merger, are very slight indeed.

If the two colliding galaxies are of very unequal size, then one is highly disturbed and the other remains intact. If a small, compact galaxy passes close to a large spiral, the spiral is relatively unaffected whereas the small compact galaxy is radically altered. If, however, the compact galaxy actually passes through the spiral, it causes the spiral to take on the shape of a ring.

The effect of galactic interactions on the gas clouds contained in galaxies is rather different. Very often the new gravitational forces acting on the clouds trigger collapses that lead to an extremely vigorous burst of star formation – a phenomenon known as a starburst. A good example is the galaxy named M82, which has been gravitationally disturbed by the large nearby spiral galaxy M81. Although the smaller galaxy has been

significantly warped, it is undergoing a rapid bout of star formation near its center.

When galaxies merge, they are stripped of the dust and gas which make new stars. Merged systems are therefore not able to generate new stars. The motions of the stars are also disturbed, so that it becomes impossible for them to settle into the ordered regime needed for a disk galaxy. The random nature of the orbits tends to make the resulting galaxies elliptical. Whether they are spherical or elliptical depends upon how random the orbits are. If the inclinations of the orbits are totally random, the galactic system is a sphere; if there is a tendency to orbital inclination, the galaxy is egg-shaped.

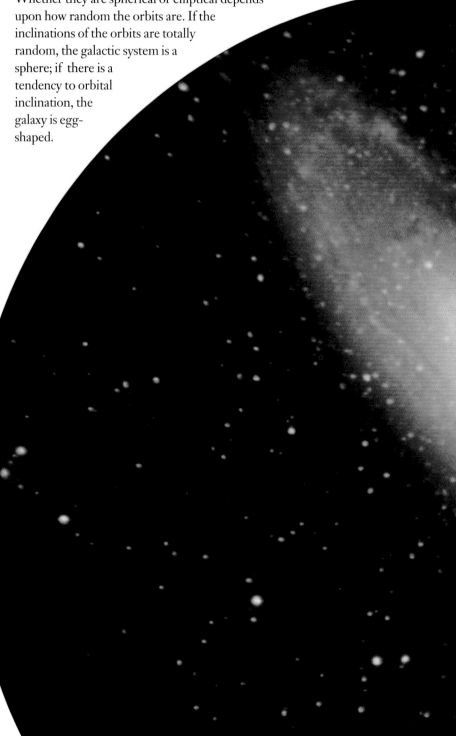

▽ A close-up view of the Andromeda elliptical galaxy shows what appears to be a double nucleus at the lower edge (usually not visible). This may be the remains of a small galaxy which was absorbed by Andromeda a billion years ago.

◁ This sequence is taken from a computer simulation that models the way galaxies collide over hundreds of millions of years. As the galaxies near one another, each begins to be distorted by the gravitational field of the other. They go into orbit around each other, gradually getting closer and closer. In the process of "spiraling" into one another, streamers of stars are thrown out behind.

4

KINDS
of Stars

STARS contribute almost all the light that makes galaxies visible. A star is a large spherical concentration of gas containing so much material that nuclear fusion takes place at its core. In fusion, light atoms combine to form heavier ones, releasing large amounts of energy. The fusion of hydrogen into helium occurs because the weight of the overlying material provides sufficient pressure to force the hydrogen nuclei close enough for fusion to occur. Not all stars generate energy in this way. For instance, if a potential star fails to gather enough matter to begin the reaction, it is called a brown dwarf star. A star may also be the hot stellar remnants of a star that has finished its phase of nuclear energy generation. Examples of these are white dwarf and neutron stars.

Stars exist in a very wide variety of sizes and luminosities. Looking up at the night sky shows how much stars vary in brightness. Some of the brightness variations seen in the night sky are due to differences in the distances of individual stars from us. The brightest stars in the sky have traditionally been recognized as patterns, which are known as a constellations. Although the brightest stars have traditional names, astronomers refer to them first by their constellation, then by assigning a Greek letter to each of its stars, starting with α (alpha) for the brightest star in the constellation.

The night sky is a jewel box of glowing stars. All of them shine with different colors and degrees of brightness. Even the naked eye can make out these differences if one stares hard enough at the stars on the celestial sphere. The variation in their appearance is a mirror of the properties of the stars themselves. Different chemical compositions, different sizes, different ages of stars – all these factors affect the way a star appears to the human observer. Understanding those differences is at the root of stellar astrophysics.

STARS AND GALAXIES

VARIOUS types of stars tend to be associated with the various different regions found in spiral galaxies (such as the Milky Way galaxy, to which the Earth belongs). Spiral galaxies consist of a nuclear bulge and a flat disk of stars. All the stars in a spiral galaxy's nuclear bulge are old. They are called population II stars and were formed when the galaxy was young. These stars are deficient in chemical elements heavier than helium (metals), which were only formed when early massive stars exploded. The orbits of the stars in the nuclear bulge have been flattened by the rotation of the galaxy. In color photographs, the nuclear region of a spiral galaxy is shown to be yellow, which is also a sign of stellar maturity.

In the disk of the galaxy there exists a collection of both young and old stars. The most prominent features in the disk, however, are the spiral arms. These outshine the rest of the stars in the disk because they contain extremely young stars – some of which are less than a million years old and very luminous. These young stars are so hot that they shine with a brilliant blue intensity which is clearly visible on color photographs. They are known as population I stars. Unlike the older, metal-deficient population II stars, they are rich in metals that enhanced the interstellar medium when much older stars exploded long ago. The Sun is an example of a typical young, hot, metal-rich Population I star.

The stars in the disk orbit the center of the galaxy. The massive blue stars that mark out the spiral arms are formed at their leading edges. This is where the interstellar medium is compressed sufficiently to collapse and form new stars. High-mass stars are short-lived, and after a few million years they explode as supernovas. This occurs toward the trailing edge of the spiral arms, and these stars do not complete a single revolution of the host galaxy.

The dimmer, low-mass stars (which are not easily visible because they are outshone by the high-mass stars) shine steadily for billions of years and circle the galactic nucleus many times. As they do so, they drift in and out of the spiral arms, unhindered by the processes that take place there.

100,000 years ago

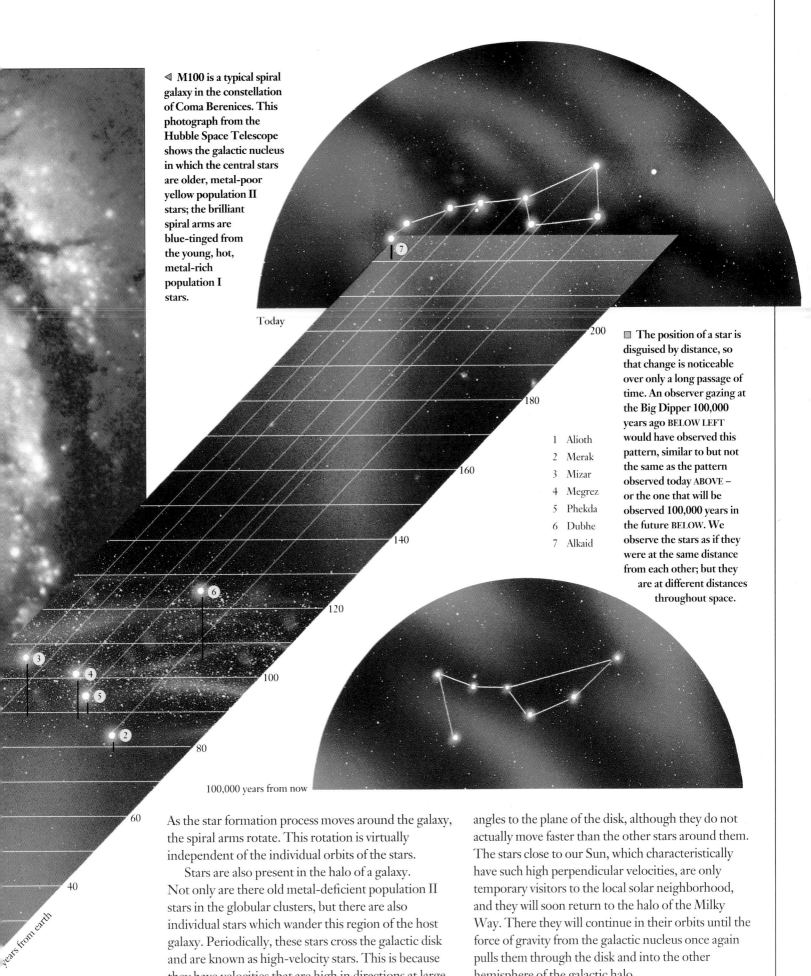

◁ **M100 is a typical spiral galaxy in the constellation of Coma Berenices. This photograph from the Hubble Space Telescope shows the galactic nucleus in which the central stars are older, metal-poor yellow population II stars; the brilliant spiral arms are blue-tinged from the young, hot, metal-rich population I stars.**

Today

200

■ The position of a star is disguised by distance, so that change is noticeable over only a long passage of time. An observer gazing at the Big Dipper 100,000 years ago BELOW LEFT would have observed this pattern, similar to but not the same as the pattern observed today ABOVE – or the one that will be observed 100,000 years in the future BELOW. We observe the stars as if they were at the same distance from each other; but they are at different distances throughout space.

180

1 Alioth
2 Merak
3 Mizar
4 Megrez
5 Phekda
6 Dubhe
7 Alkaid

160

140

120

100

80

60

40

years from earth

100,000 years from now

As the star formation process moves around the galaxy, the spiral arms rotate. This rotation is virtually independent of the individual orbits of the stars.

Stars are also present in the halo of a galaxy. Not only are there old metal-deficient population II stars in the globular clusters, but there are also individual stars which wander this region of the host galaxy. Periodically, these stars cross the galactic disk and are known as high-velocity stars. This is because they have velocities that are high in directions at large angles to the plane of the disk, although they do not actually move faster than the other stars around them. The stars close to our Sun, which characteristically have such high perpendicular velocities, are only temporary visitors to the local solar neighborhood, and they will soon return to the halo of the Milky Way. There they will continue in their orbits until the force of gravity from the galactic nucleus once again pulls them through the disk and into the other hemisphere of the galactic halo.

THE SUN

THE Earth and the other eight planets orbit a star: the Sun. It is an ordinary stellar body but looks different from the stars in the night sky because it is so close to us – 149.6 million kilometers away. It is more than one hundred times the diameter of the Earth, with nearly one third of a million times more mass.

Unlike the rocky Earth, however, the Sun is composed of 73 percent hydrogen and 25 percent helium. The remaining 2 percent is made up of the heavier elements. The Sun is a population I star – a slower-moving star found in the spiral arms of a galaxy, and believed to be relatively young.

The Sun is a fairly typical star. It has been shining for just over four and a half billion years, and will continue to do so for another four and a half billion, placing it firmly in stellar "middle age". It has an inner core (400,000 km across) in which a nuclear fusion reaction converts hydrogen into helium, accompanied by the release of vast amounts of energy in the form of heat and light. Compared with other stars throughout the Universe, the Sun is unremarkable in size or luminosity.

Being composed of gas, the Sun has no solid surface. What appears to an observer on Earth to be the visible surface of the Sun is actually a gaseous layer in which conditions promote the emission of electromagnetic radiation at visible wavelengths. Observing the Sun at other wavelengths – for example, X rays, ultraviolet and so on – allows us to see other "surfaces" to the Sun, either above or below the visible surface (known as the photosphere), depending upon the wavelength being observed. The dark atomic absorption lines in the Sun's spectrum are imposed on the Sun's light by atoms and ions in the cooler upper levels of the photosphere and in the lower part of the chromosphere, the region of gas just above it. These regions form the lowest layers of the Sun's atmosphere, above which is the more rarefied corona.

The photosphere displays a number of interesting features, most of which are driven by the electromagnetic force – one of the four fundamental forces of nature. Cooler regions of the photosphere are known as sunspots. These are produced when magnetic field lines break through the photosphere and cool the surrounding gas. Other magnetically driven

▷ **Surface activities of the Sun are the most easily visible from Earth. Sunspots are cooler regions on the photosphere and appear dark by contrast. Prominences are loops of superheated gases suspended above the photosphere on magnetic field lines. Solar flares are violent eruptions which throw vast quantities of energy and subatomic particles into space.**

▷ **The Sun is nearly 110 times the diameter of the Earth and contains most of the mass of the Solar System. This small area of the lower left quadrant is shown in the large closeup above. The Sun's visible edge (or "surface"), called the photosphere, is cool compared with the center – about 6000 K, compared with 15,000,000 K at the center and 1,000,000 K at the outer atmosphere (called the corona).**

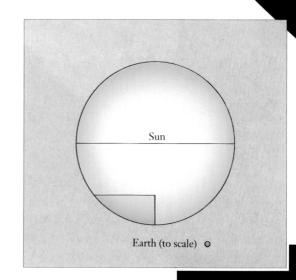

Sun

Earth (to scale) ○

phenomena are flares and prominences. Flares occur above sunspots when the energy contained in the magnetic field is suddenly released. This causes subatomic particles to be flung outward at large fractions of the speed of light and all forms of electromagnetic radiation to be spontaneously released. Prominences form when magnetic fields lift gas into the chromosphere and drape it over their field lines. Sometimes this happens over a relatively long span of time, and at other times it occurs on a timescale of about one minute.

The photosphere itself is dynamic. Huge convective cells of gas rise and fall like boiling milk and the "surface" undulates constantly. The temperature of the photosphere is estimated to be about 6000 K.

As well as electromagnetic radiation, the Sun also throws off subatomic particles in gusts known as the solar wind. The particles are accelerated into space along magnetic field lines. If these particles encounter the magnetic field of a planet, they can become trapped. When this happens to particles in the Earth's magnetic field, it causes the phenomenon called an aurora. The solar wind also causes the tails of comets.

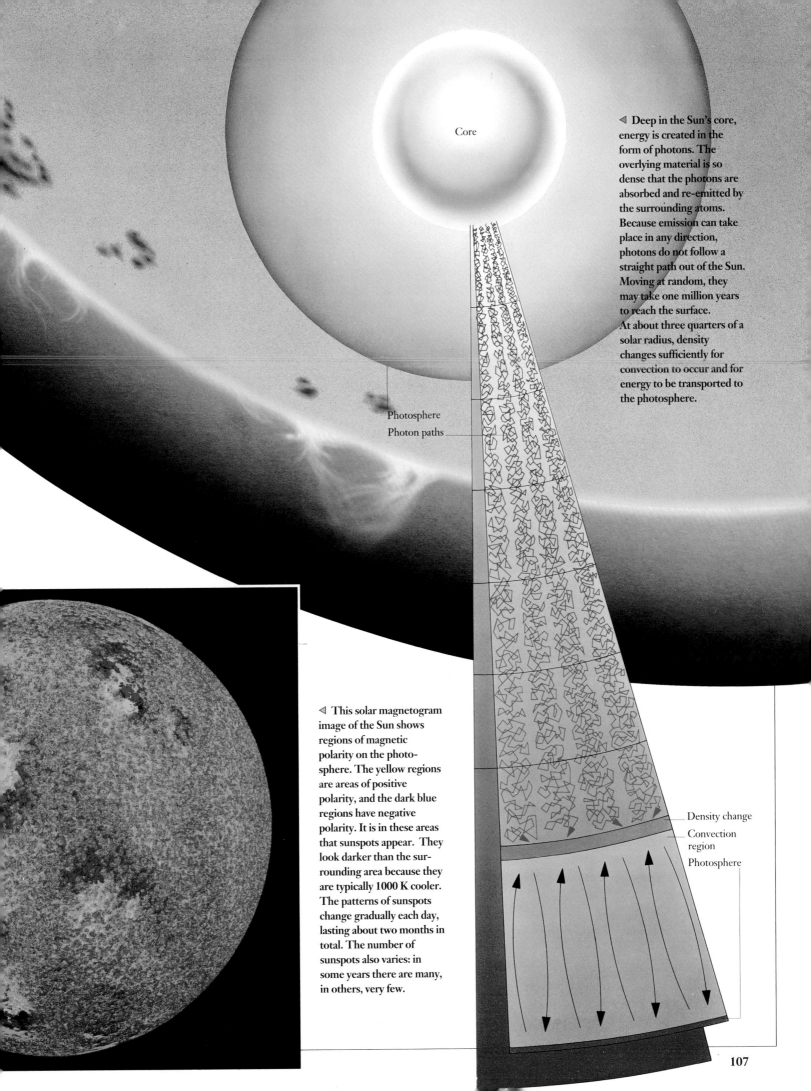

Core

Photosphere
Photon paths

Deep in the Sun's core, energy is created in the form of photons. The overlying material is so dense that the photons are absorbed and re-emitted by the surrounding atoms. Because emission can take place in any direction, photons do not follow a straight path out of the Sun. Moving at random, they may take one million years to reach the surface. At about three quarters of a solar radius, density changes sufficiently for convection to occur and for energy to be transported to the photosphere.

This solar magnetogram image of the Sun shows regions of magnetic polarity on the photosphere. The yellow regions are areas of positive polarity, and the dark blue regions have negative polarity. It is in these areas that sunspots appear. They look darker than the surrounding area because they are typically 1000 K cooler. The patterns of sunspots change gradually each day, lasting about two months in total. The number of sunspots also varies: in some years there are many, in others, very few.

Density change
Convection region
Photosphere

COLORS AND SPECTRA

EVEN a casual observer of the night sky is often able to detect variations in the colors of the stars without a telescope. One of the easiest color differences to recognize is between blue stars and red stars. For instance, the two brightest stars in the constellation Orion are Betelgeuse and Rigel. Betelgeuse is a red star whereas Rigel is white-blue. In contrast to these two extremes, the Sun is yellow.

The color of a star corresponds to the temperature of its photosphere. Red stars are cool stars of perhaps 3000 degrees whereas blue stars are extremely hot, at temperatures of 20,000 K or more. White stars are also hot stars, at around 13,000 K whereas yellow stars such as the Sun are at intermediate temperatures – only about 6000 K.

It may also be the case that the wavelength at which the most radiation is released is not a part of the visible spectrum at all. The peak emission of a 3000 degree Kelvin star actually resides in the infrared region of the spectrum. A 10,000 K star has its peak emission in the ultraviolet. The Sun emits in the middle of the visible spectrum, causing it to appear yellow.

Color is not the only characteristic determined by the temperature of a star. Temperature also determines which atomic transitions can take place within a star's atmosphere. These transitions occur when photons of light, emitted by the photosphere, are absorbed by electrons around atomic nuclei in the photosphere and lower chromosphere. These processes cause absorption lines to be superimposed upon the stellar spectrum. Spectroscopy can be used to split the light into its constituent wavelengths, so that the absorption lines can be studied. Astronomers can then use the dominant absorption lines to determine which transitions are favored and hence calculate the temperature of the star.

As well as temperature, properties such as the chemical composition, rate of rotation, density and the star's magnetic environment can be studied and the star can be classified in those terms. Each star is given a letter to designate its spectral classification. In between each classification a number between 0 and 9 is assigned to graduate the steps even further.

The letters that astronomers use to classify stars are

KEYWORDS

ATOMIC ABSORPTION
BLACK BODY RADIATION
CHROMOSPHERE
COLOR INDEX
EFFECTIVE TEMPERATURE
ELECTROMAGNETIC RADIATION
LUMINOSITY
MAGNITUDE
PHOTOMETRY
SPECTRAL CLASSIFICATION
SPECTRUM
STAR
WIEN'S LAW

▷ **Even to the naked eye, stars are noticeably different in color. This starfield contains several constellations. Orion, near the lower right, contains the red giant star Betelgeuse. Gemini is at the center left, and Taurus at the top right.**

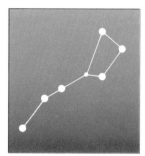

Ursa Major

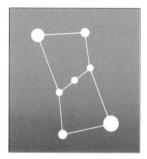

Orion

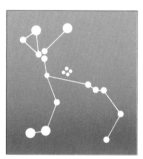

Centaurus

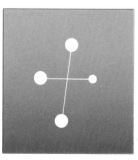

Crux

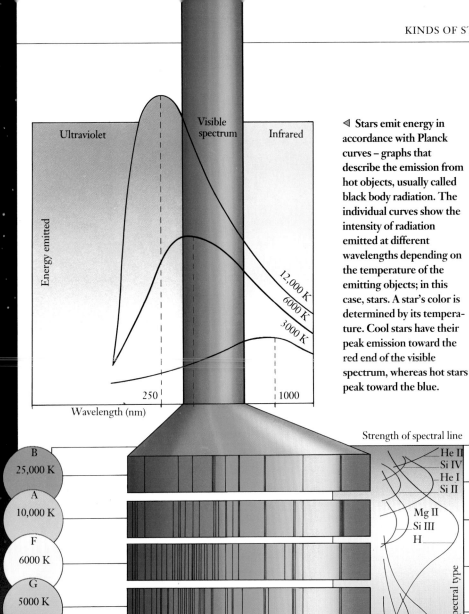

◁ Stars emit energy in accordance with Planck curves – graphs that describe the emission from hot objects, usually called black body radiation. The individual curves show the intensity of radiation emitted at different wavelengths depending on the temperature of the emitting objects; in this case, stars. A star's color is determined by its temperature. Cool stars have their peak emission toward the red end of the visible spectrum, whereas hot stars peak toward the blue.

△ Stars can be classified according to the pattern of atomic absorption lines present in their spectra. The lines are produced by electrons within atoms in the atmospheres of the stars, which absorb photons of radiation emitted at the photosphere. The temperatures in the photosphere and lower chromosphere determine the energy levels in which electrons naturally reside. This determines which absorption lines are most prominent in the spectra.

O, B, A, F, G, K and M (agreed after the original sequence of A through P, proposed in the late 19th century, had been repeatedly revised and simplified). O stars are the hottest and most massive, with temperatures in excess of 35,000 K – blue stars.

M stars are very cool, with temperatures of around 3000 K; these are the red stars. There are some additional letters used for very cool stars: R, N and S. Another spectral classification exists for a class of very hot star that periodically ejects shells of gas from its outer layers. These are known as Wolf-Rayet stars, and are given the letter W. In each of the spectral classifications there are stars of different sizes. Size also affects the brightness of a star. If two stars of different size have the same temperature, the larger star will be more luminous.

GIANTS AND DWARFS

STARS cannot be adequately classified by spectral classification alone. Although temperature is an obvious way to differentiate between stars, it does not give any indication of their size. Hydrogen-burning stars range from about one tenth the radius of the Sun to about 100 solar radii. As stars age, some grow to 1000 times the radius of the Sun. Stellar masses range from 0.08 times that of the Sun to 100 solar masses. It is unusual, however, for a star to have more than ten times the mass of the Sun.

Two stars of the same temperature but different sizes have different luminosities, which appear as subtle alterations in their spectral lines. To classify these differences, a system of five luminosity classes is used. Stars of group I are supergiants, II are bright giants, III are giants, IV are subgiants and V are known as "main sequence" stars. The main sequence classification includes late spectral type (G, K, M) dwarf stars which are burning hydrogen, such as the Sun which is classified G2 and is a yellow dwarf star. Any main sequence star with a K or M classification is a red dwarf star. White dwarfs are stellar remnants, and are not included in this scheme.

The phrase "main sequence" derives from a pattern detected when astronomers devised a single chart on which to trace the characteristics of known stars. This is known as the Hertzsprung-Russell diagram. It plots the star's luminosity against its temperature.

The Sun resides on the Hertzsprung-Russell diagram at the point where one solar luminosity crosses the temperature of 5800 K, the temperature of the Sun's photosphere. If other stars are plotted in the diagram, it soon becomes apparent that most lie in a curving S-shaped band that stretches from the low-luminosity red dwarf stars at the bottom right of the diagram up, through the Sun's position, to the high-luminosity blue stars at the top left. This is the main sequence, where stars spend the majority of their careers. The main sequence corresponds to the stable, hydrogen-burning "middle" age.

As the star ages, it begins to move off the main sequence. This is because its luminosity is caused by the energy released in its core by the nuclear fusion of hydrogen into helium. When hydrogen burning stops and helium burning begins, changes take place in the amount of energy released by the star. This internal change causes the exterior condition of the star to change as well, and it begins to change its luminosity and its temperature. Its luminosity goes up but its temperature goes down, and so it moves into the top right-hand quarter of the Hertzsprung-Russell diagram. This is the domain of the red giant stars. It is here that old stars reside in the final stages of their history.

Following the end of all nuclear reactions, most stars end up on the bottom left quarter of the diagram where white dwarfs, the stellar remnants, are plotted.

KEYWORDS

HERTZSPRUNG-RUSSELL DIAGRAM
LUMINOSITY
MAIN SEQUENCE
PHOTOSPHERE
RED DWARF
RED GIANT
SPECTRAL CLASSIFICATION
SUPERGIANT
WHITE DWARF

▷ Sirius – the brightest star in the night sky – is a brilliant white star in the constellation Canis Major (the Great Dog). At 8.7 light-years distant from Earth, it is the sixth closest star system to us. Careful observation shows that it is actually a double star with a white dwarf as a companion, with a difference in mass of 2.5 to 1. Sirius has a type A spectrum and is 26 times more luminous than the Sun.

▽ Every star in the night sky can be plotted on the Hertzsprung-Russell diagram. Devised independently by Ejnar Hertzsprung and Henry Norris Russell during the 1920s, it is a graph of the brightness of a star against its spectral classification. Most stars – stable, "middle-aged" objects such as Sirius – are found in a curving S-shaped line from the top

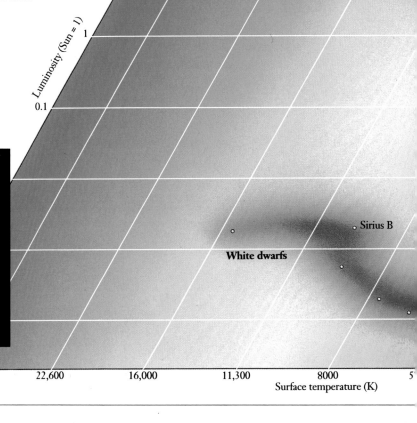

Luminosity (Sun = 1)

White dwarfs

Sirius B

Surface temperature (K)

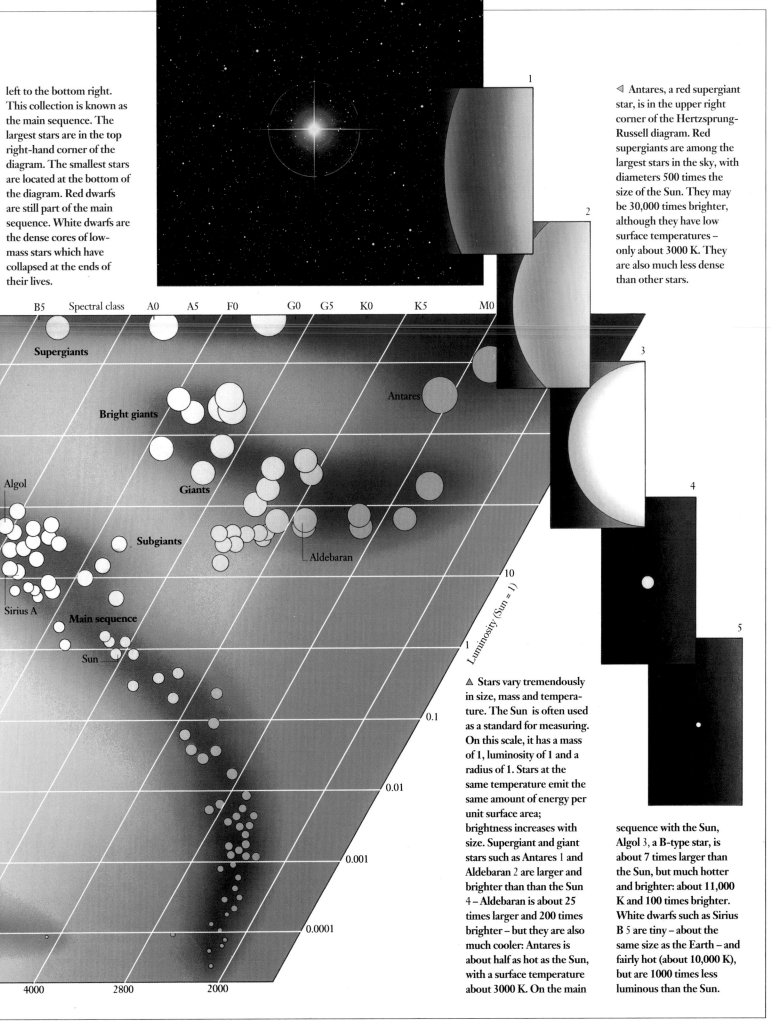

left to the bottom right. This collection is known as the main sequence. The largest stars are in the top right-hand corner of the diagram. The smallest stars are located at the bottom of the diagram. Red dwarfs are still part of the main sequence. White dwarfs are the dense cores of low-mass stars which have collapsed at the ends of their lives.

◁ Antares, a red supergiant star, is in the upper right corner of the Hertzsprung-Russell diagram. Red supergiants are among the largest stars in the sky, with diameters 500 times the size of the Sun. They may be 30,000 times brighter, although they have low surface temperatures – only about 3000 K. They are also much less dense than other stars.

Supergiants

Bright giants

Antares

Giants

Algol

Subgiants

Aldebaran

Sirius A

Main sequence

Sun

Luminosity (Sun = 1)

10

1

0.1

0.01

0.001

0.0001

Spectral class

B5   A0   A5   F0   G0   G5   K0   K5   M0

4000   2800   2000

△ Stars vary tremendously in size, mass and temperature. The Sun is often used as a standard for measuring. On this scale, it has a mass of 1, luminosity of 1 and a radius of 1. Stars at the same temperature emit the same amount of energy per unit surface area; brightness increases with size. Supergiant and giant stars such as Antares 1 and Aldebaran 2 are larger and brighter than than the Sun 4 – Aldebaran is about 25 times larger and 200 times brighter – but they are also much cooler: Antares is about half as hot as the Sun, with a surface temperature about 3000 K. On the main

sequence with the Sun, Algol 3, a B-type star, is about 7 times larger than the Sun, but much hotter and brighter: about 11,000 K and 100 times brighter. White dwarfs such as Sirius B 5 are tiny – about the same size as the Earth – and fairly hot (about 10,000 K), but are 1000 times less luminous than the Sun.

BINARY AND MULTIPLE STARS

MOST stars do not exist in isolation; they have stellar companions and orbit each other. Occasionally it is possible to see the two components, either with the naked eye or through a telescope. In such a case the star is called a visual binary. Not all stars that look close together are true binary stars, however. Some stars are not associated with each other and are located at vastly different distances, but because they lie in the same direction from the Earth, they look close together in the sky. True binaries are ones in which two stars are bound together by the force of gravity. They may have begun life as two protostars, or as a single protostar that split apart.

The time taken for binary stars to orbit each other is highly variable. It depends upon many factors, such as the mass of the two components, the ratio of their individual masses, how far apart they are and what stage in their evolution they have reached. Some stars spin around one another in a few days, whereas others take centuries.

Many double stars cannot be seen as visual binaries. Perhaps the star system is too far away for the separate components to be resolved, or perhaps it is relatively near but the two components are too close together. Sometimes one component is much fainter than the other and is outshone as a result.

The components of a binary star system orbit each other around their common center of mass – neither star is stationary. If this oscillating motion can be detected in relation to the background of stars, it indicates that a small or dim companion star is in orbit around the larger, brighter one. Such pairs are known as astrometric binaries.

Another method of finding binaries is to study their spectra. The spectral absorption lines may indicate the presence of two stars, each with a different spectral classification. Even if the stars are exactly the same type, they are in motion and this causes the spectral lines to alter in wavelength. This is because the wavelength of radiation emitted from a moving object is either stretched or squashed, depending upon whether the object is approaching or receding – this phenomenon is known as the Doppler effect. The stars are moving in different directions, causing the spectral lines to alter their wavelengths by different amounts. As a result, in the course of a single orbit the spectral lines move apart and come back together

KEYWORDS

ASTROMETRIC BINARY
BINARY STAR
DOPPLER EFFECT
DOUBLE STARS
MASS
ORBIT
SPECTRAL ABSORPTION
  LINES
SPECTRAL
  CLASSIFICATION
VISUAL BINARY
WAVELENGTH

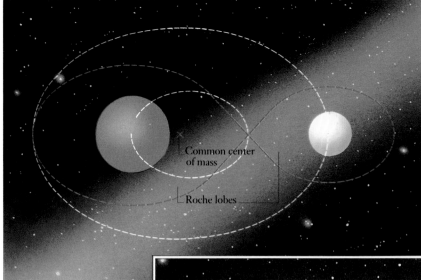

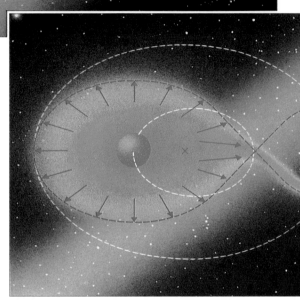

▷ **In the second stage of a binary star's lifetime, the more massive of the two stars has become a red giant and filled its Roche lobe. The material which is ejected from this star via a stellar wind, crosses to the gravitational field of the smaller star, is captured and spirals down on to its surface. This process begins to increase the smaller star's mass.**

▷ **In the third stage, the red giant star completes its evolutionary cycle. At this point it becomes either a white dwarf or a neutron star. The once-smaller companion star, which is still on the main sequence, continues to evolve, but it now does so at a much faster rate because of the increased mass of the material which has accreted onto its surface from the former red giant star.**

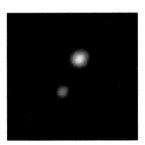

Mizar

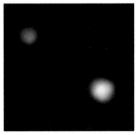

β Cygni

γ Andromeda

◁ The three photographs of binary stars illustrate three different binary star systems. Among these stars, the widest separation is 34 arcseconds and occurs in the beta Cygni star system, some 410 light years from Earth. The closest system is gamma Andromeda with a separation of only 10 arc-seconds. Different colors indicate different masses or different stages in the stars' evolutionary cycles. Mizar, or zeta Ursa Majoris, is a triple star.

◁ For much of their lives, binary stars only influence each other in terms of their orbits. In the first stage, the two stars revolve around their common center of mass. The boundary of each star's gravitational field is known as a Roche lobe. At the point where these lobes touch their gravitational pulls cancel out each other.

Light

Time behind large star

Time in front of large star

Time

△ A binary star system is known as an eclipsing binary if the two companions are arranged so that they cause mutual eclipses, as viewed from Earth. This causes variations in the light output from the system. The stars are brightest when one is visible alongside the other; the greatest drop in brightness occurs when the brighter star is eclipsed by the dimmer one, even if the brighter star is the larger of the pair. The motions of the two stars can usually be distinguished by spectroscopic analysis. The best-known eclipsing binary is Algol, or beta Persei; its magnitude varies between 2.2 and 3.7 over a period of under three days.

twice. If the companion star is too faint, its spectrum is swamped by that of the brighter star. That spectrum still exhibits the Doppler shift, however, and the presence of a companion can be inferred from that. This system is called a spectroscopic binary.

Binary star systems provide an opportunity for astronomers to weigh stars. To do this, the distance between the stars and the time it takes for them to complete an orbit must be measured. Using simple mathematics, a figure for the combined mass of the two stars can then be calculated. An estimate as to which star contains most mass must then be made. If the two stars are identical, however, the figure can simply be halved.

Triple star systems are also known, as are quadruple stars and systems with even more components. The more stars contained in a multiple system, the rarer they are. More than half of all the known stars exist in binary systems or in systems with as many as six members.

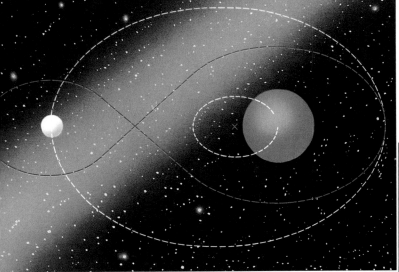

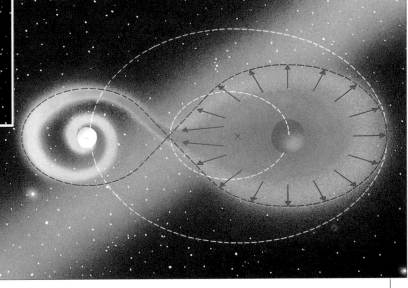

▷ In stage four, the companion star (once the smaller star) has finally become a red giant. Like its companion star before it, the second star swells up to many times its normal size. It, too, fills its Roche lobe and begins transferring mass back to the first star. What happens next depends on what the first star has become. If the transferred material falls onto a white dwarf, a nova will occur; if it falls onto a neutron star, an X-ray burst will occur.

VARIABLE STARS

I F A double star has its orbital plane inclined edge-on to the Earth, the two stars periodically eclipse each other and one star blocks the light of its companion. This causes the system to vary in brightness, depending on the relative luminosities of the stars involved, whether the eclipses are total or partial, whether the stars are gravitationally distorting each other, and whether they are transferring mass and producing hot, bright spots. A graph of the brightness variation is known as a light curve. It takes various shapes, depending on the different conditions.

Some individual stars also vary in brightness. This indicates the internal processes that occur in the variable star. There are two broad types of variables, pulsating and eruptive. Pulsating variables alter their brightness in a regular, periodic way, accompanied by pulsations of the stellar "surface" in which the star actually expands and contracts. Cepheids are yellow giant stars that pulsate in very regular ways. They are very bright and easily recognizable, and are often used to determine the distances to nearby galaxies. There are two types of Cepheids. Type II Cepheids, also known as W Virginis stars, are less luminous than type I. Red giant stars are also prone to variation and these are known as Mira-type variables. The star Mira varies by up to a factor of 10,000 over a period of 80 to 1000 days. Other red giant stars vary gently and fairly unpredictably over timescales of several years.

Eruptive variables include faint red dwarf stars, known as flare stars. These are cool and faint and, at first, seem very unlikely candidates for an unpredictable rise in luminosity. In fact, what may be happening is that they undergo flares in the same way as the Sun but, because the star is so much fainter than our own, the flare has a much more dramatic effect upon its appearance.

Novas are stars that suddenly flare in brightness to many thousands of times their former luminosity. They are thought to be binary star systems in which matter from the larger star is being transferred onto a white dwarf companion. That material then ignites in a runaway nuclear fusion process and the star flares in brightness. As material builds up again another nova is possible. If the material builds up sufficiently, the white dwarf is destroyed in the nuclear detonation, releasing much more energy. The system is then totally destroyed in a type I supernova – an explosion so powerful that it can become as luminous as a galaxy for a few days after the explosion. Type II supernovas are not as bright and occur when a high-mass star reaches the end of its life and blows itself to bits.

Most of the variable activity of stars derives from the fact that the star is aging. Before becoming a main sequence star, however, young stars undergo intense periods of unpredictable variability. This occurs because it takes a while for the stars to settle down to the nuclear fusion processes which have just ignited in their cores. In stars of similar mass to the Sun this is known as the T Tauri phase.

KEYWORDS

BINARY STAR
CEPHEID VARIABLE STAR
DOUBLE STAR
ECLIPSING BINARY
FLARE STAR
MIRA-TYPE VARIABLE
NOVA
RED DWARF
RED GIANT
SUPERNOVA
T TAURI PHASE
VARIABLE STAR

▽ **A nova is thought to occur in a binary star system in which one member becomes a white dwarf. Initially both stars are on the main sequence but one is more massive than the other. This star evolves faster and becomes a red giant. It becomes so large that it fills its Roche lobe, within which all material is dominated by the star's gravity. Material crossing this limit is transferred from the larger star to the small one. By the next stage, the larger star has died and left a white dwarf in its place. The mass transferred to the small star has accelerated its evolution and caused it to**

1 Binary system
2 Large star becomes red giant
3 Red giant dies and becomes white dwarf
4 Small star becomes red giant
5 Material transfers to white dwarf
6 Material compresses
7 Nuclear fusion begins
8 Nova

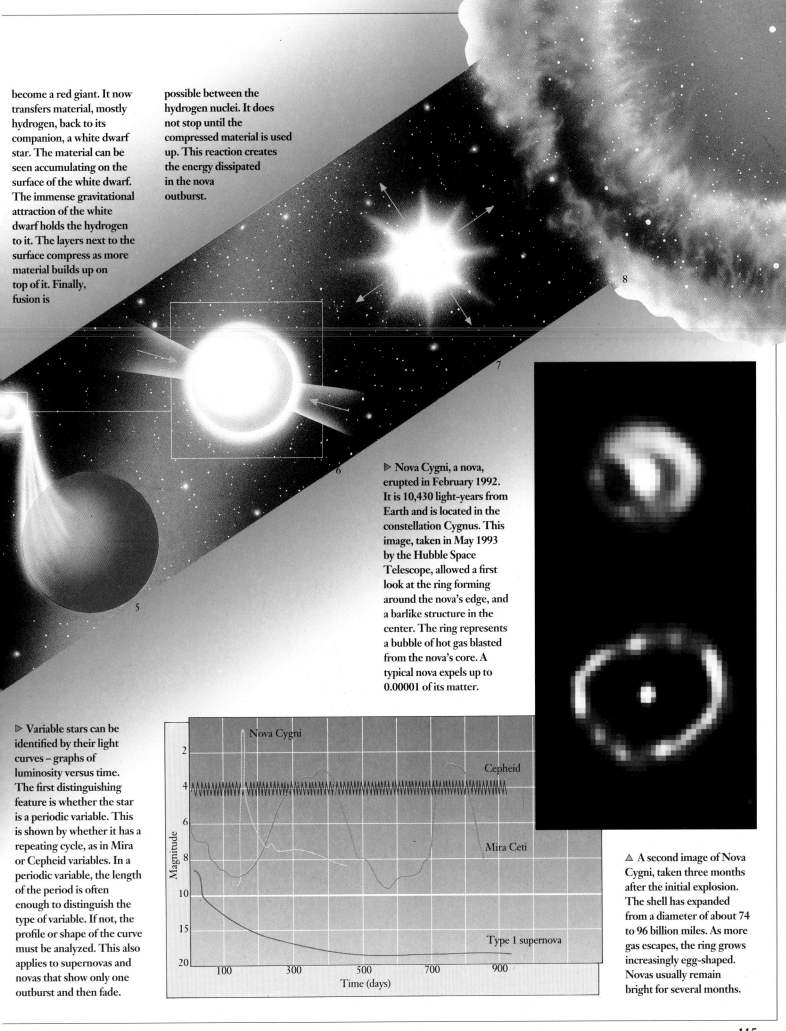

become a red giant. It now transfers material, mostly hydrogen, back to its companion, a white dwarf star. The material can be seen accumulating on the surface of the white dwarf. The immense gravitational attraction of the white dwarf holds the hydrogen to it. The layers next to the surface compress as more material builds up on top of it. Finally, fusion is

possible between the hydrogen nuclei. It does not stop until the compressed material is used up. This reaction creates the energy dissipated in the nova outburst.

▷ Nova Cygni, a nova, erupted in February 1992. It is 10,430 light-years from Earth and is located in the constellation Cygnus. This image, taken in May 1993 by the Hubble Space Telescope, allowed a first look at the ring forming around the nova's edge, and a barlike structure in the center. The ring represents a bubble of hot gas blasted from the nova's core. A typical nova expels up to 0.00001 of its matter.

▷ Variable stars can be identified by their light curves – graphs of luminosity versus time. The first distinguishing feature is whether the star is a periodic variable. This is shown by whether it has a repeating cycle, as in Mira or Cepheid variables. In a periodic variable, the length of the period is often enough to distinguish the type of variable. If not, the profile or shape of the curve must be analyzed. This also applies to supernovas and novas that show only one outburst and then fade.

△ A second image of Nova Cygni, taken three months after the initial explosion. The shell has expanded from a diameter of about 74 to 96 billion miles. As more gas escapes, the ring grows increasingly egg-shaped. Novas usually remain bright for several months.

LIFE & DEATH
of a Star

5

ASTRONOMERS have studied so many stars that they have
been able to build up a detailed picture of how they believe
these fascinating objects "live" and "die". Stars change
considerably over time – but this time is often measurable only in
millions of years. For this reason, astronomers never actually see a
star move from one phase of its life to another, much less witness a
complete lifespan. However, there are so many stars in the sky that it
is possible to observe each of the individual stages by looking at
different stars. From the observations astronomers make, they must
try to deduce the correct sequence of events in a star's evolution and
identify where each star is placed along the sequence.

Stars form within vast clouds of dust and gas. Complex chemical
reactions take place within these giant molecular clouds. Although
the vast majority involves molecular hydrogen, the small amounts of
heavier elements present also take part in the reactions and form
more complex molecules. A particularly fascinating discovery is that
giant molecular clouds contain organic (carbon-containing)
compounds. These compounds are necessary to the development of
life. Astronomers are currently searching for amino acids and sugar
molecules. If they prove to be abundant, it may give a clue as to how
life arises in the Universe.

Stars do not occur uniformly throughout space, but are confined to galaxies. Even within galaxies, though, there are only certain regions in which star formation takes place. These sites are areas in which dust and gas have accumulated to a greater density than elsewhere in space. The constellation of Orion contains a vast star-forming region of which the Horsehead nebula is a small part. The shape is a random distribution of dust which absorbs the light from the glowing hydrogen gas behind it. Orion, which is relatively close to Earth, is one of the most thoroughly studied sites of star formation.

BIRTH OF A STAR

As GIANT molecular clouds orbit the center of a galaxy, they are tugged by gravitational and magnetic fields. How fast their constituent particles move depends on their temperature: the colder the cloud, the slower the particles. Fast-moving particles resist collapsing together, and so stars can form only in the dense cores of cold clouds. Typically, these clouds are only about 15 degrees above absolute zero. Periodically, the clouds begin to collapse. The trigger mechanisms for such collapses are thought to be collisions between giant molecular clouds or entry into galactic spiral arms.

Both of these occurrences set up compression waves within the cloud, which cause isolated regions to become so dense that gravity overwhelms all other processes and the cloud collapses. These isolated regions can often contain enough mass to create several hundred stars of similar mass to the Sun. They are known as Barnard objects, and often appear as black regions in front of stars. Sometimes regions with emission nebulas reach the appropriate density and collapse. These appear as round, black "bubbles" within the glowing gas. They are referred to as Bok globules. As Barnard objects and Bok globules collapse, isolated regions within them collapse as well. In this way, the cloud fragments on many different scales. It is the smaller-scale collapses from which stars form.

At the center of the collapsing regions, concentrations of matter build up. Three-quarters of this matter is in the form of hydrogen gas. The rest is nearly all helium with 2 percent being made up of the heavier elements. This region is known as a protostar and, as material pours down upon it, the gas becomes so compressed that the temperature begins to rise dramatically. The rise in temperature makes the gas move faster and thus creates more pressure. This pressure gradually balances the inward pull of gravity and halts the collapse of the protostar. As more material accumulates on the protostar, instead of collapsing, it is squeezed gently. This raises the temperature still further.

Although there are no nuclear processes going on within the protostar, it is still giving off energy from the material that is striking its surface. This is given off as radiation but is very quickly absorbed by the dusty

KEYWORDS

ABSOLUTE ZERO
BARNARD OBJECT
BECKLIN-NEUGEBAUER OBJECT
BOK GLOBULE
DENSE CORE
EMISSION NEBULA
GIANT MOLECULAR CLOUD
GRAVITY
HYDROGEN
INFRARED
PROTOSTAR

▷ The Hubble Space Telescope took this high-resolution, false-color image of the Orion star-forming region. It emits hydrogen, ionized oxygen and sulfur, which are shown as green, blue and red respectively. At the upper center of the image is a jet of material being shot out by a young star. This material is clearing out a cavity which will become a reflection nebula. Only one jet can be seen because the other is pointing away and has been obscured by dust. There are many of these objects in Orion's nebula.

△ Very young stars are often found at the centers of bipolar nebulas, which may form when subatomic particles and radiation from the young star carve out shapes from the interstellar medium. 1 The density of matter in the center of the collapsing region builds up by accretion. The impact of material falling onto the central protostar heats the object, and energy is given off. Energy is also produced by the nuclear fusion of the hydrogen isotope deuterium at a much lower temperature than normal hydrogen fusion. Deuterium burning may help create the protostellar wind which carves out the bipolar cavities. In stage 2, the bipolar nebula begins to take on a characteristic shape, and an accretion disk has developed around the protostar. This concentration of dust acts as a barrier that stops radiation and subatomic particles escaping along the equatorial plane of the young stellar object. In the polar regions, the density of material is minimal; radiation escapes from these. 3 The nebula is now mature and readily detectable. Photons escape from the protostar and travel through the cavities. When they hit the cavity walls they scatter in all directions. Some of them are projected toward Earth. By studying the polarization of this light, astronomers can deduce much information about the central star.

envelope raining down upon the surface of the protostar. This action heats the dust, which then re-radiates the energy at infrared wavelengths. The envelope that surrounds a protostar is vast; typically, it is 20 times larger than our entire solar system.

The first, young, infrared star to be found was discovered in the Orion star-forming region. It was discovered in 1967 by Eric Becklin and Gerry Neugebauer of the Californian Institute of Technology, and is now known as the Becklin-Neugebauer object. The youngest protostar, however, is in the constellation of Ophiuchus and is known as VLA 1623, named for the Very Large Array telescope from which it was discovered. It is thought to be less than 10,000 years old.

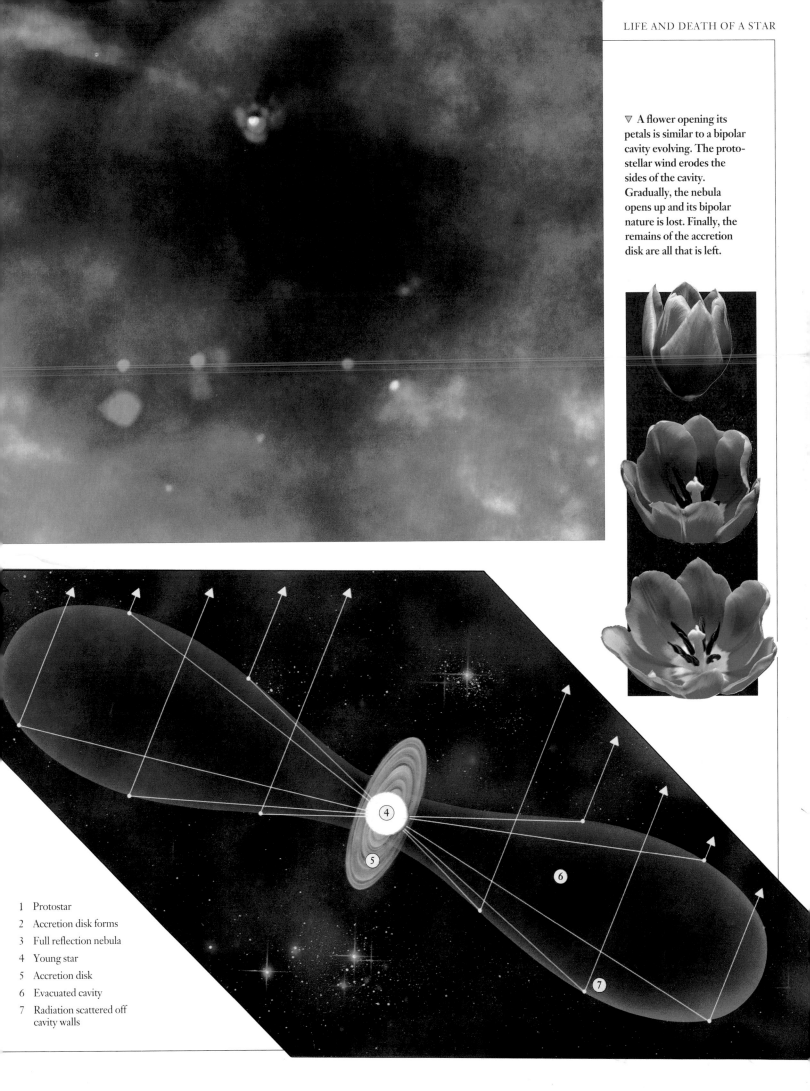

▽ A flower opening its petals is similar to a bipolar cavity evolving. The proto-stellar wind erodes the sides of the cavity. Gradually, the nebula opens up and its bipolar nature is lost. Finally, the remains of the accretion disk are all that is left.

1 Protostar

2 Accretion disk forms

3 Full reflection nebula

4 Young star

5 Accretion disk

6 Evacuated cavity

7 Radiation scattered off cavity walls

FORMATION OF PLANETS

A S GRAVITY causes molecular clouds in space to collapse onto a protostar, the material does not simply fall straight onto the protostar. Instead, because of the cloud's original rotation, it swirls downwards and forms a thick disk of material around the star. This is known as an accretion disk.

The protostar emits subatomic particles such as electrons and protons. This emission forms a stellar wind which pours out of the protostar. It is very difficult for this wind to move through the accretion disk because of the large amounts of obscuring material there.

However, along the disk axis there are regions with relatively small amounts of infalling dust and the particles can readily carve a path away from the protostar. The cavities thus created are conical in shape and, once they have been cleared of dust, allow radiation from the protostar to escape through them as well. When that radiation hits the cavity walls, it is scattered by the dust there. Radiation that is scattered towards Earth is visible to us, and for this reason the object is called a reflection nebula. The cavities, because they extend from the rotation poles of the protostar, are known as bipolar outflows.

The accretion disk is the location where astronomers believe that planets form. Although planets have not yet been found in orbit around stars other than the Sun, this is probably because even modern telescopes are still too weak to detect them. Dusty disks of material have been found around stars such as Vega and Beta Pictoris. These may be sites of planet formation or even reservoirs of comets, waiting to fall in towards the central star.

There are two theories that explain how planets are formed in accretion disks around stars. In the first theory, the disk develops gravitational instabilities in the same way that the parent gas cloud fragmented and collapsed. These instabilities collapse under gravity, dragging material in with them, and become protoplanets. They eventually finish collapsing and planets are formed as a result.

The other theory is known as collisional accretion. Here, the dust particles in the accretion disk collide with one another. Some particles stick together and thereby grow a little in size. This increases their mass

KEYWORDS

ACCRETION DISK
ASTEROID
BIPOLAR OUTFLOW
COMET
EARTH
ELECTRON
GRAVITY
NUCLEAR FUSION
PLANET
PROTON
PROTOSTAR
REFLECTION NEBULA
STELLAR WIND
SUBATOMIC PARTICLE

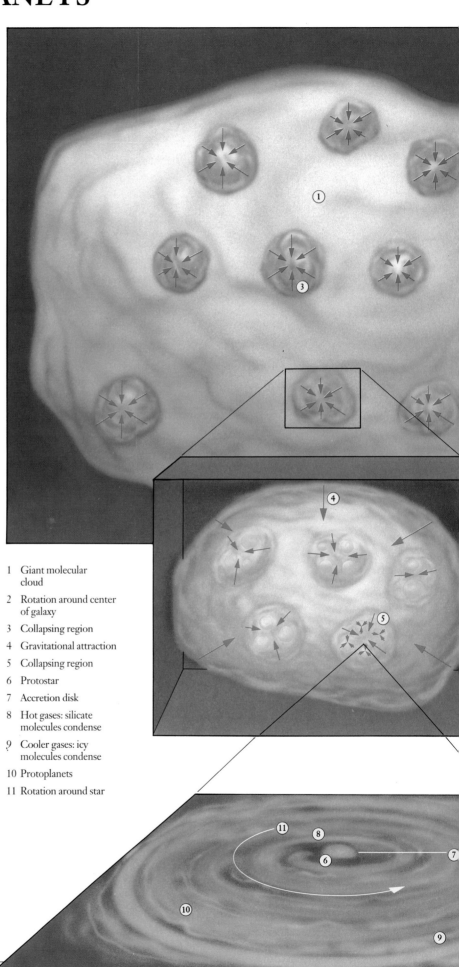

1 Giant molecular cloud
2 Rotation around center of galaxy
3 Collapsing region
4 Gravitational attraction
5 Collapsing region
6 Protostar
7 Accretion disk
8 Hot gases: silicate molecules condense
9 Cooler gases: icy molecules condense
10 Protoplanets
11 Rotation around star

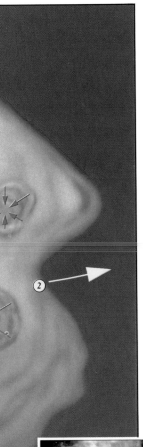

◁ The formation of planetary systems around stars occurs at the same time as the star itself forms. In the first stage, the giant molecular cloud – which is in orbit around the center of the galaxy – develops gravitational instabilities in areas of high density 1. These collapsing regions are known as dense cores. Stage 2 shows the internal processes of the dense core. They look remarkably like those of the whole cloud, with further collapsing regions. One of these regions may develop into a nebula, with a dense core at the center, which will become the star, surrounded by a hot, spinning disk of gases, which cool and condense to form particles that eventually collect together (accrete) to form planets.

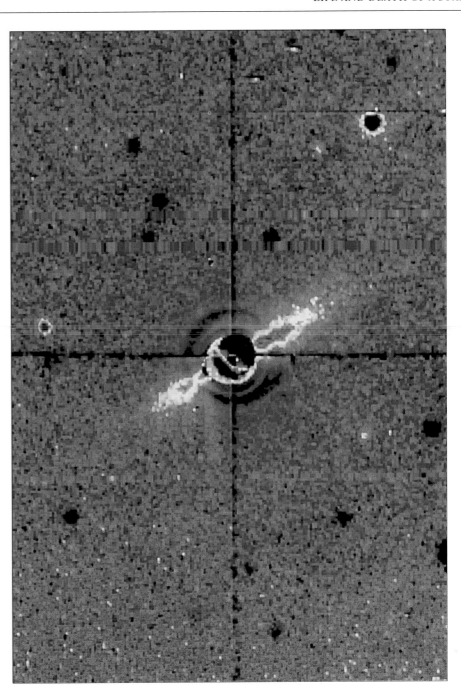

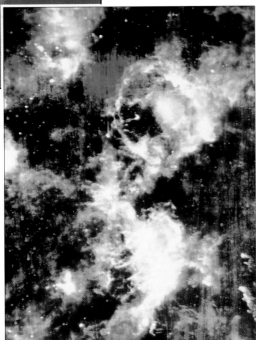

△ The giant molecular cloud in the constellation Orion is so large that, if our eyes could see it, it would cover the whole constel-lation. The large bright area is the nebula, a major star-forming region.

and hence their gravitational pull. When they collide with another particle of dust, it adheres to them and, gradually, over thousands of years, rocky particles known as planetesimals are formed.

Planetesimals are similar to the present-day asteroids. They continue to orbit the central star but, since their orbits cross one another, they frequently collide. This causes them to conglomerate until planet-sized bodies are formed. The Solar System is thought to have begun with a large number of planet-esimals, which gradually gathered together over 30 million years to form protoplanets, and finally, after 150 million years, formed the four inner planets.

After perhaps a million years, planet formation is stopped when the parent star reaches the point in its evolution at which nuclear fusion begins in its core – about 8 million degrees.

△ The star Beta Pictoris possesses a circumstellar disk of material. The star is 50 light-years away and the disk can be clearly seen in this image, in which the star has been blocked out so that it does not outshine the fainter, reflected light from the disk. The disk extends for 60 billion kilometers away from the star – ten times farther than the orbit of Pluto in our Solar System. Instead of being a solar system in formation, it is possible that the disk around Beta Pictoris is a Kuiper belt, a reservoir of comets.

ON THE MAIN SEQUENCE

As PROTOSTARS accumulate more and more mass, the temperature and pressure in their centers grow ever higher until they force protons close enough for nuclear fusion to take place. This begins the processes by which the star generates energy, and the protostar begins to join the main sequence on the Hertzsprung-Russell diagram.

Before it can settle into stable "middle age", however, it must adjust to the nuclear processes now taking place within its core. These processes provide a pressure that pushes material outward. Gravity, which has governed the formation of the star so far, is eventually equalized by the pressure of the hot gas inside the star, and the star stops collapsing.

While pressure equalization is taking place, the star undergoes a dramatic and unpredictable variability in its luminosity and outflow of material. This behavior can also excite small regions of the surrounding molecular cloud, causing them to emit radiation. These emission regions, with a characteristic knotty appearance, are known as Herbig-Haro objects.

The masses of the stars formed within the collapsing fragments depend upon factors such as the mass of material contained in the fragments and the rate at which the material accretes onto the protostars. In any collapsing cloud, the stars formed may range in size from the largest to the smallest known stellar masses.

In general, the less massive stars are produced in much greater numbers than the high-mass stars. Along the spectral classification sequence OBAFGKM, the most abundant stars are the red dwarfs with spectral types of K and M. The high-mass, high-luminosity, short-lived stars of O and B type are very few in number, but these are highly important to the evolution of the star-forming region. These stars are prodigious in their release of radiation and generate intense stellar winds – subatomic particles – which are accelerated along magnetic field lines away from the star. The radiation ionizes hydrogen in the surrounding envelope, creating free electrons and protons. When these particles reform into hydrogen atoms, they give off electromagnetic radiation. One of the common wavelengths for emission is in the red

region of the optical spectrum and so these emission nebulas, often called HII regions, glow with a characteristic red color. As well as ionizing their surroundings, these high-mass O and B stars also push the surrounding material away from them. This causes the molecular cloud in their vicinity to be compressed, and therefore starts the process of collapse in new regions. In this way, the process of star formation propagates throughout a giant molecular cloud, with regions of newer star formation associated with older regions. These are known as OB Associations.

The mass of a star determines just how long it spends on the main sequence. The more massive stars use their fuel at such vast rates that they only have enough hydrogen to last a few tens of millions of years. The lower-mass stars, despite having less hydrogen to fuse, exist on the main sequence for much longer because they use that hydrogen at a much more leisurely pace. A G-type star, such as the Sun, fuses hydrogen into helium for about nine billion years. A red dwarf star, which uses hydrogen very slowly, will continue to exist on the main sequence for many tens of billions of years.

When stars emerge from their birth clouds, they are usually in associations known as star clusters. A good example of this is the young star cluster known as the Pleiades (Seven Sisters). As these clusters orbit the galactic center, the individual stars gradually drift apart and, eventually, lose their association with one another. In the case of the Sun, for example, it is now impossible for scientists to determine which were the stars with which it formed. Analysis has shown, however, that most of the stars that form the constellation of the Big Dipper are associated with one another and formed at the same time.

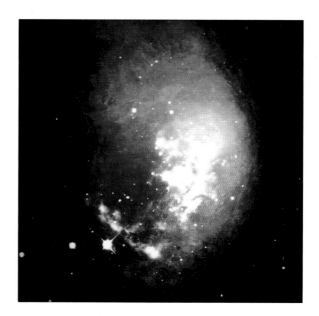

◀ Infrared wavelengths can penetrate the dust clouds in which stars form. This view of Orion is a composite of several infra-red wavelengths with the colors indicating temper-atures. Reddish regions are at 15 K, running up to violet at 100 K. The four bright stars right of center are O and B stars.

△ As time goes by, stellar winds from the O and B type stars blow away the surrounding material. The Rosette Nebula is a cloud of dust and gas 100 light-years across. The material blown outward by the central stars has caused the other regions to begin collapsing. Thus, dark Bok globules surround the central cavity.

▷ After millions of years, stars become completely dissociated from their birth clouds. The Pleiades star cluster (M45) is a group of stars which all formed together from the same molecular cloud. All that is now left of their natal cloud are the wispy filaments which can be seen around a few of the stars.

POST-MAIN SEQUENCE

THROUGHOUT its entire main-sequence lifetime, only about 10 percent of the hydrogen in a star takes part in nuclear fusion. As the core runs out of hydrogen, fusion is confined to a shell around the inert core of helium. Because there are no energy-generating mechanisms at work in the core, it begins to contract, causing higher temperatures and pressure.

Core contraction is transmitted through the release of potential energy to the outer layers of the star. As these layers are bloated outward, the star puffs up greatly in size – often becoming 10–100 times larger in diameter. The temperature drops, however, and the star becomes redder. The star is now called a red giant.

As the core continues to shrink, the temperature continues to rise. Eventually, the temperature rises high enough so that helium fusion begins in the core. The ignition is known as the helium flash, and the nuclear reaction employed is known as the triple alpha process. This involves two helium nuclei, often referred to as alpha particles, fusing together to form a radioactive beryllium isotope. If a third helium nucleus collides before the beryllium can decay, a stable carbon nucleus is produced. Sometimes, a fourth alpha particle takes part in the reaction and an oxygen nucleus is formed.

Carbon is produced until the helium is used up, and again the core becomes inert. In stars of similar mass to the Sun, this is as far as the nuclear fusion process goes. As the nuclear fusion processes subside and finally stop, the star begins to die. In these low-mass stars, the outer gaseous layers are ejected by stellar wind processes. These may be the inevitable result of the pulsations which giant stars often suffer as they become Cepheid variable stars. The result of the mass loss, however, is to produce a planetary nebula. As the

KEYWORDS

BLACK DWARF

CEPHEID VARIABLE STAR

ELECTRON DEGENERATE MATTER

MAIN SEQUENCE

PAULI'S EXCLUSION PRINCIPLE

PLANETARY NEBULA

RED DWARF

RED GIANT

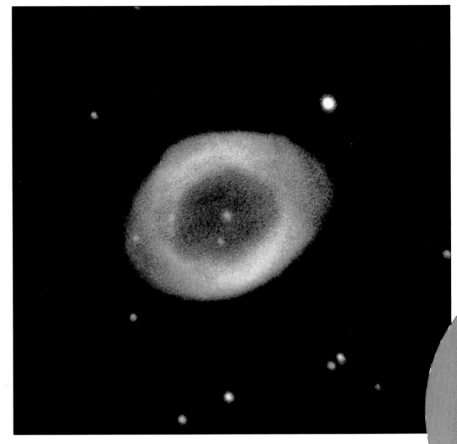

▽ **The evolutionary sequence of a star with a similar mass to the Sun** begins with the creation of the star from a collapsing cloud of gas, and follows it through the main sequence to the onset of helium burning, the red giant stage and beyond.

1 Protostar
2 Main sequence phase
3 Expansion phase
4 Red giant
5 Contracting phase
6 Planetary nebula
7 White dwarf

▷ **The outward appearance of a star depends on what is happening in its core.** When the star is burning hydrogen into helium, it remains on the main sequence. When the hydrogen is exhausted and the conditions promote the fusion of helium into carbon, the star "puffs up" into a red giant. In a low-mass star, such as the Sun, the end of helium burning results in a collapse, which results in the formation of a planetary nebula and a white dwarf.

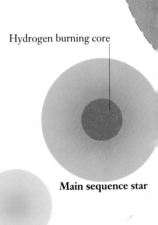

Hydrogen burning core

Main sequence star

Protostar

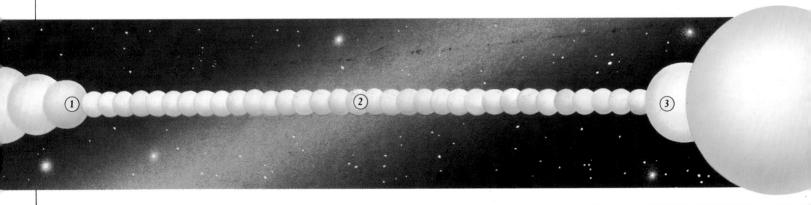

White dwarf

Planetary nebula

◁ **The ring nebula M57, in the constellation of Lyra,** is a classic example of a planetary nebula – a thin, expanding cloud of gas blown into space at the end of a star's red giant phase. The red emission comes from hydrogen, the green from oxygen and the yellow from neon.

Gas shell

Stellar wind

Hydrogen burning shell

Inert helium core

Pulsating red giant

Expanding envelope

Hydrogen burning shell

Inert helium core

▷ **The triple alpha reaction is believed to be the** predominant method of energy generation in the cores of red giant stars. It is a way of synthesizing carbon from three helium nuclei, sometimes referred to as alpha particles. In the first stage, two helium nuclei fuse to form a nucleus of beryllium. This, in turn, collides and fuses with another helium nucleus to form a carbon nucleus. The reaction takes places at temperatures greater than 100 million kelvin.

Helium

Helium

Helium

Helium

Beryllium

Beryllium

Proton

Neutron

Carbon

Gamma ray

Gamma ray

gas drifts outward from the central star, it is illuminated by the radiation that is still being released by the star. This radiation is absorbed by the gas in the shell and re-emitted at visible wavelengths. The nebula gradually disperses and merges into the interstellar medium after a few hundred thousand years.

The central stars of planetary nebulas have been pulled into small, compact objects by the force of gravity. Sometimes they are white dwarfs – stars in which the density of matter is so great that the electrons can no longer exist in orbits around the nuclei. Instead, they are all compressed so that they try to occupy the lowest energy levels.

This is a state of matter known as electron-degenerate matter and results in the electrons exerting a great pressure because, as dictated by Pauli's exclusion principle, they cannot occupy the same quantum states as one another. This pressure, known as electron-degeneracy pressure, stops the white dwarf from collapsing any more.

The white dwarf cools gradually over the course of the next few billion years, until it finally becomes a body known as a black dwarf. The largest stars, though, suffer a much more dramatic fate and can leave even more compact objects than white dwarfs.

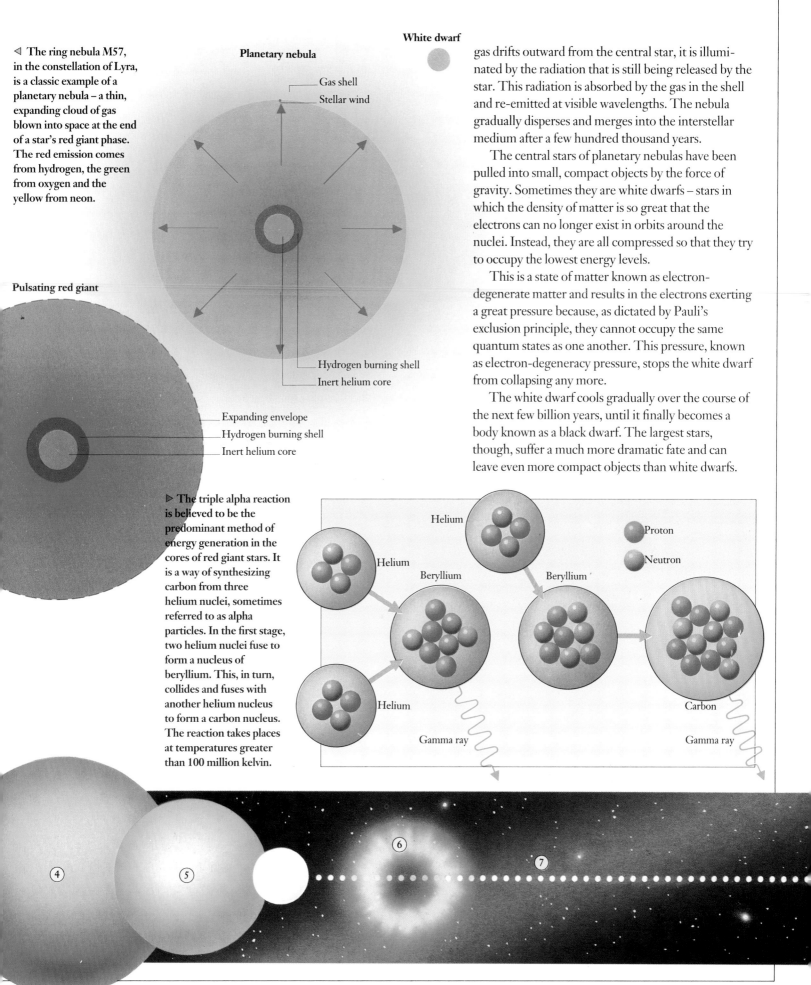

4

5

6

7

COLLAPSING AND EXPLODING STARS

I N STARS of more than seven solar masses, the inert carbon core is so massive that it collapses sufficiently to ignite carbon fusion. The temperature needed to ignite the carbon is in the region of several hundred million degrees. The carbon fusion produces magnesium. The star begins to take on a layered structure, with each shell within the center of the star undergoing nuclear fusion of a different element. Hydrogen fusion still takes place in the outermost shell of the core region and below that, helium is converted to carbon and oxygen. The star develops concentric rings of material, each of which is fusing a specific chemical element into another one and feeding the shell below it. The shells contain such elements as neon, sodium, magnesium, silicon, sulphur, nickel, cobalt and, finally, iron. These high-mass stars race through their final evolutionary phases at extraordinary speed, compared with their initial phases (some of which lasted for tens of thousands of years, and some for millions of years). Carbon fusion is usually complete within a thousand years, neon and oxygen fusion takes place within a single year. The silicon burning, which produces the iron core, usually takes place within a mere day or two.

Iron builds up in the core of the star and does not fuse into anything else. This is because all of the nuclear fusion processes so far have released energy but, after iron, the energy needed to fuse elements together is greater than the energy released in the fusion process. There is nowhere for that energy to come from, and so the iron accumulates in an electron-degenerate mass at the center of the star. The electron pressure is not infinite, however, and, as the mass gradually builds up, the core begins to become unstable. When the mass contained in the core reaches just under one-and-a-half times the mass of the Sun, known as the Chandrasekhar limit (after the Indian-US astrophysicist Subrahmanyan Chandrasekhar), the electron pressure can no longer resist the pull of gravity and the core collapses even further. This collapse has the same effect as knocking the foundations out from beneath a building. The overlying structure, in this case the rest of the star, begins to collapse downward.

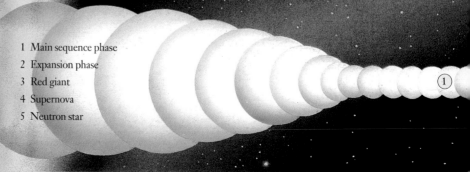

1 Main sequence phase

2 Expansion phase

3 Red giant

4 Supernova

5 Neutron star

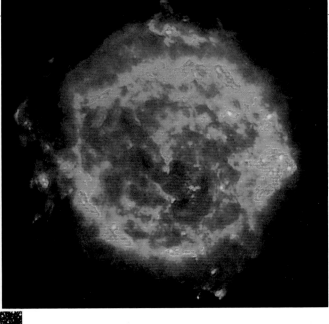

As the star crashes down upon itself, it releases so much energy that it explodes and virtually blows itself to bits. This is known as a supernova. The energy released in supernova explosions initiates the production of the elements heavier than iron. Stars that explode in this way are called supernovas, type II. Type I supernovas involve white dwarf stars. If a white dwarf star is close enough to another star, that star can transfer some of its outer atmosphere onto the white dwarf. This builds up on the white dwarf until a catastrophic nuclear detonation takes place. This can destroy the white dwarf and produce a supernova type I.

Supernovas seed the interstellar medium with elements that are heavier than helium. The Universe is composed of 75 percent hydrogen and 23 percent helium; heavier elements make up the remaining two percent. Those heavier elements, called metals by astronomers, make planets and life possible. Every atom on Earth and in our bodies was once at the center of a massive star which exploded as a supernova before the Sun and the planets formed. The shock waves from supernova explosions are one of the mechanisms by which the interstellar medium is compressed and thus new stars formed.

◁ The Vela supernova remnant is a roughly circular shell of luminous gas filaments. This photograph shows only part of the shell. The star that gave rise to this nebula is believed to have gone supernova some 12,000 years ago. Almost at the center of the shell is pulsar 0933-45, which is thought to be the remains of the exploded star.

△ This is a view of supernova remnant Cassiopeia A taken at a radio wavelength of 6 cm. It exploded about 300 years ago and is one of the brightest radio sources in the sky. The areas in blue indicate the most intense emission.

▽ When a star of more than 1.4 solar masses (the Chandrasekhar limit) leaves the main sequence, it expands to form a red giant. Eventually it explodes as a violent supernova and blasts its outer layers of material into space. The core collapses under gravity to form a tiny, extremely dense neutron star. Luminosity increases by a factor of 10^8 when a star goes supernova; this lasts only a few days.

Iron, nickel, cobalt
Silicon
Oxygen
Carbon
Helium

◁ As a massive star reaches the end of its life, fusion processes occur in the layers of its core, possibly simultaneously. Hydrogen fusion gives helium, helium fusion gives carbon and oxygen, carbon fusion gives neon and magnesium, and oxygen fusion gives silicon and sulfur. Finally, silicon and sulfur fusion gives iron. Iron fusion requires supplying additional energy, so the iron accumulates as an inert core.

2   3   4   5

NEUTRON STARS AND PULSARS

WHEN the core of a massive star can no longer withstand the pressure caused by the downward pull of gravity, the stellar material collapses into a state called degenerate matter. Degenerate matter is matter in which the normal arrangement of atoms has broken down under the force of gravity because the weight of overlying material is so oppressive. In baryon degenerate matter, the electrons – which usually orbit the atomic nuclei – have been forced into the nuclei, where they combine with the protons and form neutrons. As a result, the entire core of the star is composed of neutrons, tightly compressed.

Under these conditions the neutrons are still being tugged by gravity. But, according to Pauli's exclusion principle, in spite of the densely crowded conditions, no two identical particles may occupy the same quantum state (a set of conditions of location, spin and and velocity that may apply to a particle). In other words, two neutrons cannot be in the same place at the same time; it is physically impossible. So, just as the electrons did before them, the neutrons exert a pressure which resists further collapse, which would pull them even closer together.

Objects composed of neutrons are extremely compact. White dwarf stars, composed of electron-degenerate material, typically have a diameter similar to that of the Earth. They do, however, contain more mass than the Sun. Neutron stars are even more extreme: they contain more mass than one-and-a-half Suns packed into a spherical region with a diameter of only 10 to 20 kilometers – equivalent to the nuclear density of an atom. The star's ability to resist gravitational collapse by means of degeneracy pressure is limited by its mass. Up to 1.4 times the mass of the Sun, the star can support itself by electron degeneracy pressure; this is called the Chandrasekhar limit. Beyond 1.4 solar masses, the matter collapses until it is halted by baryon degenerate pressure, which is effective up to the Oppenheimer-Volkoff limit of between three and five solar masses. This is the upper mass limit for a neutron star.

Neutron stars are left behind following supernova type II explosions. They are the collapsed cores of massive stars. Although their existence was predicted by theory in the 1930s, it was thought that they would be undetectable because of their small size. Then, in the 1960s, a class of rapidly pulsating objects which became known as pulsars was discovered. It was soon shown that the only objects that could have such behavior were spinning neutron stars. They operate in a similar way to a lighthouse: although the light appears to flash on and off, this is an illusion produced by a rotating light. In this same way, the beam of radiation is swept through space by a pulsar. As it crosses our line of sight, we receive a pulse of radiation. Astronomers do not yet understand how this radiation is produced, nor why it is confined in such narrow beams.

After a supernova explosion, the neutron star is left spinning at high velocities. For instance, the pulsar in the center of the Crab nebula – which is the remains of a star that went supernova in AD1054 – is spinning so fast that it flashes 30 times a second. The fastest pulsars are known as millisecond pulsars, and they can spin at hundreds of revolutions per second. These are old pulsars which have been "spun up" by accretion from nearby stars. This "spinning up" process is similar to the way in which material is funneled onto white dwarfs in binary star systems. Because of the properties of degenerate matter, the more mass a neutron star accumulates, the smaller it shrinks. The smaller a neutron star becomes, the faster it spins. Its magnetic field also increases by a factor of one billion (in proportion to its compressed surface area).

The process of spinning up a neutron star leads astrophysicists to expect an accretion disk to be formed around it. Around protostars, the accretion disk is the site of planet formation. A tentative discovery suggests that two planets have, indeed, formed around a pulsar, known as PSR1257+12. However, these "second generation" planets are not locations on which it is conceivable that life could develop.

▷ A pulsar is thought to flash on and off because it is rotating. This is known as the lighthouse model, because a lighthouse uses a rotating mask with an aperture in it to create the same kind of flashing. Observation of pulsars shows that the pulses are emitted at many different wavelengths. Although the pulse profiles are slightly different, the pulses all occur at the same time. Pulses may vary in speed between different pulsars; some are fast, such as the pulses in a close binary star, and some slow.

1 Core
2 Neutron fluid
3 Solid crust
4 Axis of rotation
5 Charged particles
6 Magnetic field lines
7 Pulsar emission

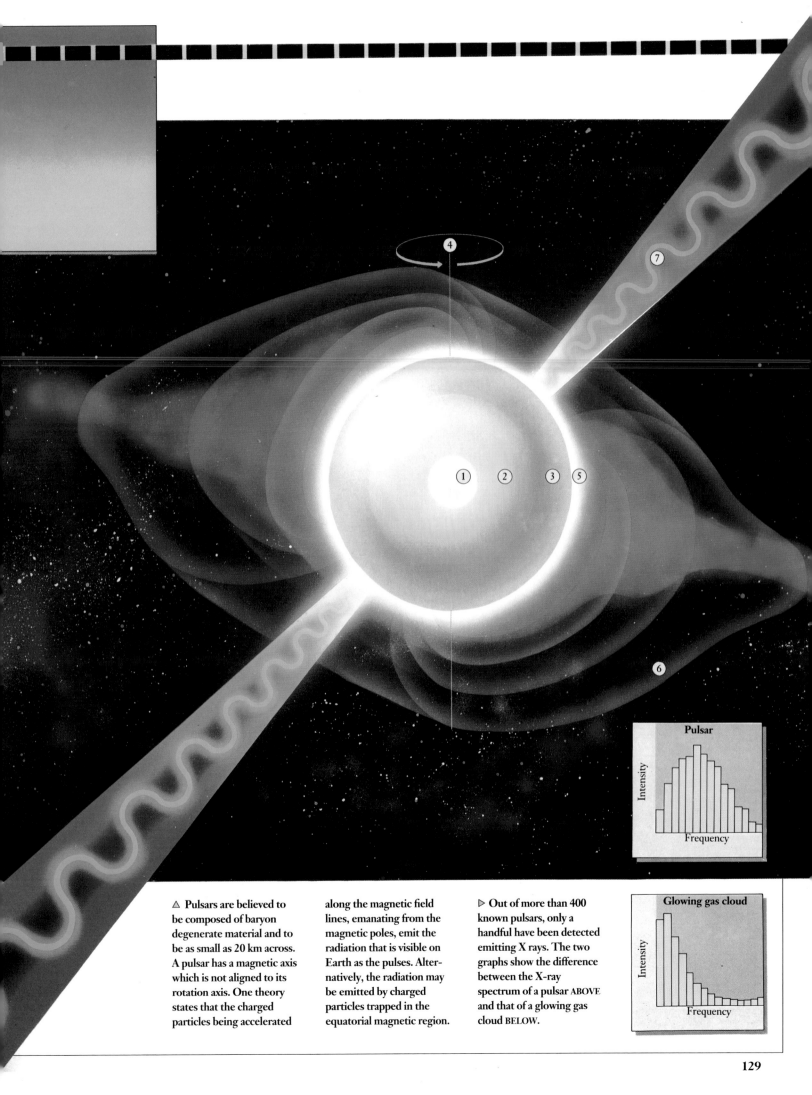

Pulsar

Intensity / Frequency

Glowing gas cloud

Intensity / Frequency

△ Pulsars are believed to be composed of baryon degenerate material and to be as small as 20 km across. A pulsar has a magnetic axis which is not aligned to its rotation axis. One theory states that the charged particles being accelerated along the magnetic field lines, emanating from the magnetic poles, emit the radiation that is visible on Earth as the pulses. Alternatively, the radiation may be emitted by charged particles trapped in the equatorial magnetic region.

▷ Out of more than 400 known pulsars, only a handful have been detected emitting X rays. The two graphs show the difference between the X-ray spectrum of a pulsar ABOVE and that of a glowing gas cloud BELOW.

BLACK HOLES

A COLLAPSAR is a star that has collapsed because nuclear fusion is no longer taking place within its core. Some collapsars become white dwarfs and neutron stars; others become black holes. The mass of the star determines whether it becomes a white dwarf, a neutron star or a black hole.

Black holes are objects that have exceeded a calculable limit for inert, non-energy producing material, about 3.2 times the mass of the Sun. Above this limit – called the Oppenheimer-Volkoff limit – the baryon degeneracy pressure, which causes neutrons to exert pressure to resist the force of gravity, can no longer halt the collapse of the star due to gravity. The smaller the star becomes, the greater the force of gravity at its surface. The greater the surface gravity, the greater the velocity that would be needed to escape from the star. As the collapsing star gets smaller and smaller, the escape velocity rises until it is equal to the speed of light. When the escape velocity reaches this level, nothing can escape from the star – not even light – and it becomes a black hole. The star has then disappeared from the observable universe, although some of its effects can still be detected.

If the star is very massive, it requires less compression to become a black hole. The radius within which a star must be compressed before becoming a black hole is called the Schwarzschild radius. It defines the region known as the event horizon – the edge of the black hole, where the escape velocity equals the speed of light. No one can ever know what takes place inside the boundaries of the event horizon. It is thought that the collapsing mass continues to shrink until it becomes a minuscule point of infinite density known as a singularity.

The gravity around a black hole is so strong that it causes space to "curve" around it. Astronomers believe that black holes rotate, causing them to drag the spacetime continuum in their immediate vicinity around with them. This region of space is known as the ergosphere. Its edges are marked by the stationary limit, inside which nothing can be held stationary but is dragged around by the black hole. Objects crossing the event horizon are lost forever and are believed to disappear into the singularity.

Because nothing can escape from a black hole, it is very difficult to discover one. Like white dwarf and neutron stars, black holes can exist in binary star systems. Gas from the companion star is stripped by the gravitational influence of the black hole and funneled down into it. Because the star and the black hole rotate around one another, the material forms an accretion disk around the black hole. The material in this disk swirls around the black hole so fast that the friction between the molecules heats the gas until it begins to emit X rays; as it does, it loses energy and spirals into the black hole. The X rays can be detected on Earth, indicating a black hole.

An example of a possible black hole is Cygnus X-1, an X-ray source in the constellation of Cygnus which orbits a blue supergiant star of between 20 and 30 solar masses. This massive star seems to be pulled gravitationally by an invisible companion with between 9 and 11 solar masses. The X-ray emission is thought to come from an accretion disk around the companion. Supermassive black holes, thousands of times the mass of the Sun, are thought to exist at the centers of active galaxies and quasars.

▷ The black hole Cygnus X-1 orbits a blue supergiant star which it is slowly ripping apart. The outer layers of the star travel down onto the black hole, spiraling into an accretion disk, which is so hot that it emits radiation as X rays. These can be detected from the Earth.

1 Supergiant star
2 Black hole
3 Accretion disk
4 Hot spot

◁ The central white region of M87, an elliptical galaxy in the Virgo cluster, displays a density of stars 300 times greater than in a normal elliptical galaxy. Astronomers believe that the stars are held there by a black hole of 2.6 billion solar masses. The effects of the black hole are limited to a very small region of space, only a few million kilometers around the hole.

△ A black hole acts as a gravitational lens when it comes between an object and an observer. With a simulated hole in front of a picture of Albert Einstein ABOVE LEFT, the alignment is not perfect; most of the image is on one side, with a thin arc on the other CENTER. With perfect alignment, the image RIGHT becomes a ring, known as an Einstein ring.

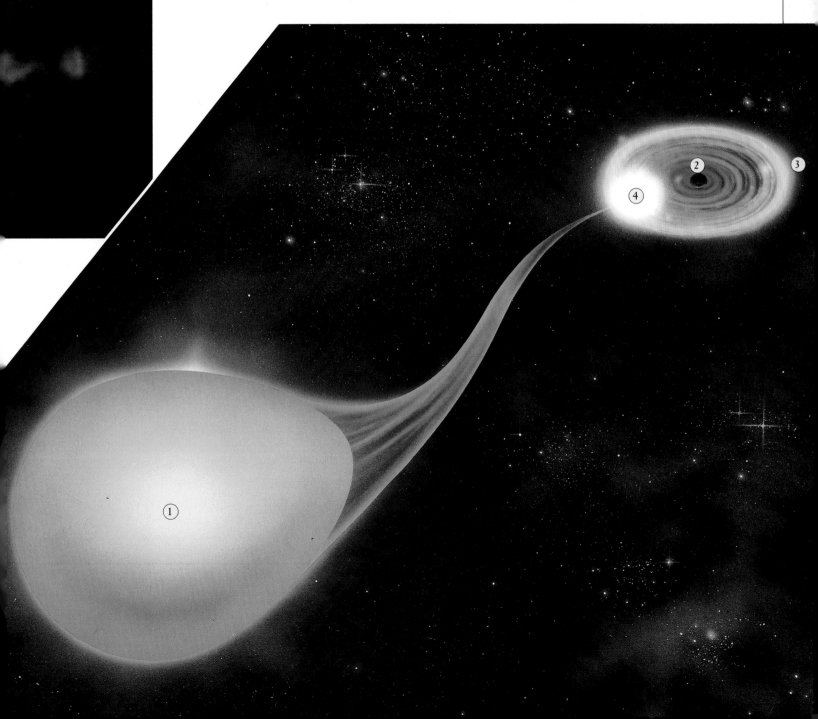

6

THE FATE
of the Universe

EVERY KNOWN OBJECT in the Universe is made of protons, neutrons and electrons. Everything we see is visible because it emits or reflects photons of electromagnetic radiation. Are there other forms of matter that we cannot detect?

Astronomers now suspect that other forms not only exist, but heavily outweigh normal, so-called baryonic, matter. Studies of galactic clusters suggest that the amount of matter in such clusters may exceed luminous material by ten to one. This means that 90 percent of the Universe is contained in forms of matter that are yet to be discovered – with profound effects on the question of whether the Universe will one day collapse under the force of gravity.

This invisible material is called dark matter. Two possible forms have been suggested: hot and cold. Hot dark matter may comprise particles such as neutrinos which are extremely lightweight, travel at the speed of light and barely interact with baryonic matter. Cold dark matter may comprise the hypothetical particles sometimes referred to as Weakly Interacting Massive Particles (WIMPS). An alternative theory states that dark matter is normal baryonic material, held in nonluminous objects such as brown dwarfs and black holes, which may exist in galactic halos and make up entities known as MAssive Compact Halo Objects (MACHOS).

As well as dealing with the way in which the Universe came into being, cosmology also deals with the way in which it will develop in the future. Ultimately, cosmology must predict the way in which the Universe will end. Our predictions of the fate of our Universe depend primarily how much matter remains to be discovered in the Universe. The US Hubble Space Telescope is playing a pivotal role in trying to detect what is called "dark matter".

OPEN, FLAT OR CLOSED?

THE question of how much mass the Universe contains has a direct bearing on its eventual fate. It has long been known that the Universe is expanding. Astronomers are now asking, will it ever stop expanding? The answer depends on how much mass there is in the Universe, and hence the total force of gravity within it.

The presence of mass curves the spacetime continuum. On the very largest scale, the curvature of the Universe is determined by the average density of matter within it – that is the average mass contained within a specific volume of space. The average density required to halt the expansion of the Universe – known as the critical density – is only a few hydrogen atoms per cubic meter. The ratio of the average density of the Universe to the critical density is known as Ω (omega).

A Universe with Ω less than 1 will exist and expand forever and is known as an "open" Universe; its spacetime continuum has what astronomers call negative curvature. A Universe within which the expansion is halted through force of gravity is known as a "closed" Universe and spacetime has positive curvature. A third possibility exists, known as the "flat" Universe. This occurs if there is just enough matter to halt the expansion, but only after an infinite period of time. Current estimates suggest that the average density of the Universe may equal the critical density. If so, the Universe is "flat" and will exist forever.

KEYWORDS

GRAVITY

INFLATIONARY
  COSMOLOGY

MASS

SPACE

SPACETIME CONTINUUM

▷ Although astronomers have reliable ways of calculating the amount of matter in a star, or even in a galaxy, it is not so easy to weigh all the matter in the entire Universe. Instead, astronomers look at the effect of the curvature of the Universe on our view of distant galaxies. If space were positively curved through force of gravity, we would expect parallel lines to converge eventually, and therefore that the density of galaxies would decline the further away we look. In fact, studies of the deep sky (as shown in this photograph) show a consistent distribution of galaxies, which suggests that space has an even geometry. Studies of the density of very distant galaxies also support this conclusion: if the Universe were closed, we could expect a decrease in the density of distant galaxies.

◁ The geometry of a closed Universe is shown here on the hemisphere and the distorted image of Albert Einstein (who himself did not believe the Universe was expanding). On a sphere, parallel lines converge. If a standard image of Einstein is projected onto a sphere, then replotted onto a flat plane (like our view of the celestial sphere), the extremities of the face are stretched, and the center squeezed. This supports the notion that, in a closed Universe, distant galaxies would appear less densely packed than closer ones.

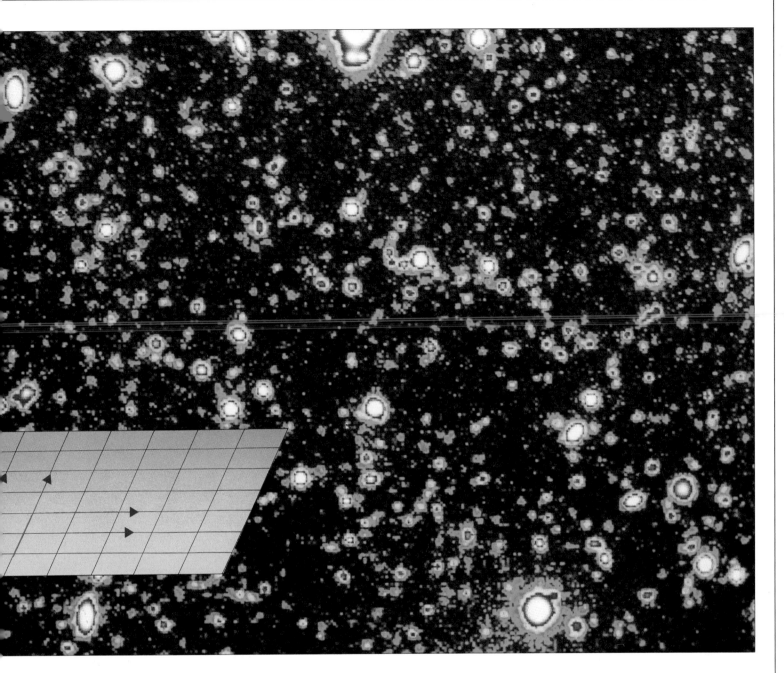

△ In a flat Universe, parallel lines would remain parallel for ever, and an even distribution of matter such as galaxies through-out space would appear to us as just what it was. This hypothetical state of affairs is confirmed by the image of Einstein: in a flat geometry, no distortion appears. This geometry is supported by studies of deep space, and by implication is that the Universe is not closed, after all, and therefore will always exist; the galaxies will continue to move apart, gradually slowing but never coming to a complete stop.

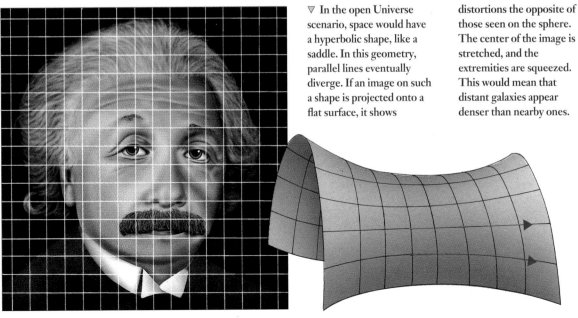

▽ In the open Universe scenario, space would have a hyperbolic shape, like a saddle. In this geometry, parallel lines eventually diverge. If an image on such a shape is projected onto a flat surface, it shows distortions the opposite of those seen on the sphere. The center of the image is stretched, and the extremities are squeezed. This would mean that distant galaxies appear denser than nearby ones.

THE LONG-TERM FUTURE

IF THE Universe is "flat" or "open", it will exist for an infinite length of time. This does not mean, however, that planets, stars and galaxies will also exist forever. The Universe is governed by the laws of physics. One of these laws, the second law of thermodynamics, states that heat flows from a hot object to a cold one. So that when both objects are at the same temperature, the flow of heat stops; it can never flow from a cold object to a hot one. Every chemical process that takes place in the Universe follows this guiding principle. In effect, the stars and galaxies slowly lose heat to the surrounding Universe, and so die.

Before this happens, more and more stars in galaxies will come into close proximity with one another. This will result in one star being ejected from the galaxy while the other plunges toward its central regions. The matter in the center of galaxies will become more and more compact and, eventually, black holes with the mass of galaxies will form. This same process will be repeated in clusters of galaxies because some galaxies will be ejected and others will fall toward the central regions. Thus the Universe will be filled with black holes that contain the same amount of mass as galaxy clusters.

The matter contained within these black holes will be re-processed and returned to the Universe via the Hawking radiation process. This is a process by which a pair of virtual particles is created right on a black hole's event horizon. One of the particles escapes, whereas the other falls in, negating some of the mass of the black hole. It appears as if the escaping particle has come from the black hole itself, which gradually "evaporates" into space. The smaller the black hole, the faster it evaporates. This evaporation can be measured as heat. As particles escape and the mass of the black hole decreases, its temperature goes up. The increased temperature causes more particles to escape, further decreasing the mass and raising the temperature. Finally, in the last seconds, the black hole releases the remainder of its mass in a cataclysmic burst of energy equivalent to the explosion of one billion one-megaton hydrogen bombs.

By this process – stars coalescing into black holes which then evaporate – given a long enough period of time, the entire contents of the Universe will reach

thermal equilibrium. When this happens there will be no stars, planets or galaxies, only a tenuous "sea" of subatomic particles. All particles will be at the same temperature and nothing will react. If chemical reactions are no longer taking place in this Universe, there will be nothing by which to judge the passage of time. The Universe will have died. This concept is known as heat death.

If, though, the Universe is "closed", then expansion will eventually slow and stop, and it will begin to collapse. Clusters and then individual galaxies will merge together. The cosmic microwave background radiation will increase its temperature and eventually space will become so hot that stars will evaporate. The Universe will have returned to conditions very similar to those that existed during the Big Bang. Instead of expanding, however, the Universe will be shrinking and heading towards a Big Crunch.

Some astronomers have suggested that the Big Crunch will so precisely match the conditions of the Big Bang that the Universe will be reborn. That new Universe may be very different from ours, however, because the laws of physics themselves would have to be melded all over again in the first few fleeting instances of expansion.

▷ An "open" universe has too little matter to create enough gravitational field to halt the expansion of space. An open universe will therefore expand for ever. Although the expansion will be slowed by the gravitational attraction of the matter it contains, it will never be halted and reversed. The Universe will suffer from "heat death", which occurs when all the objects in it have become the same temperature. The likely timescale for this is 10^{12} years. At 10^{30} years, after the remains of dead galaxies have formed supergalactic black holes, protons may begin to decay into electrons and positrons; all matter will then do the same.

1

△ The amount of matter in the Universe determines the way in which the spacetime continuum is curved, and hence the future of the Universe. At present, many observations suggest that the Universe is "flat". The chances of a totally flat universe are so improbable, however, that these observations have become known as the flatness problem. A refinement of the theory of the Big Bang has been proposed in order to explain it. It is known as inflation, and it supposes that shortly after the Big Bang the Universe was driven to inflate by exponential proportions. Thus, whatever the true curvature of the Universe, it will always appear flat to us. This is similar to the way in which the Earth appears flat to us even though it is a sphere.

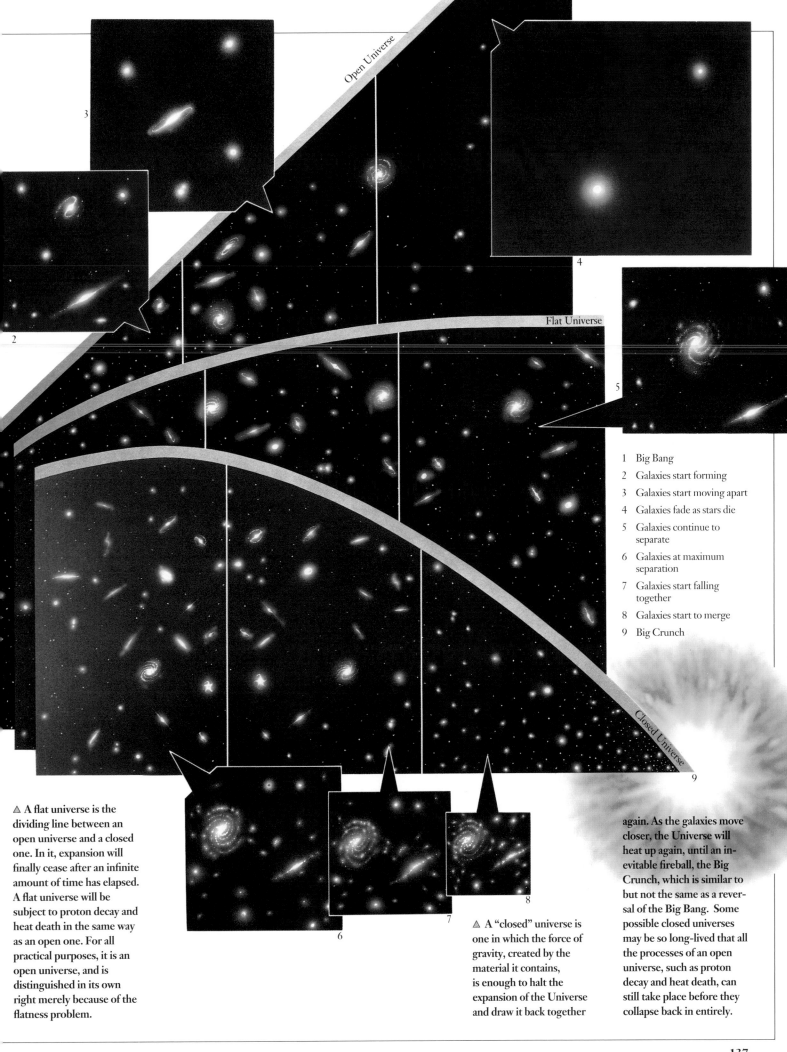

Flat Universe

Closed Universe

1   Big Bang

2   Galaxies start forming

3   Galaxies start moving apart

4   Galaxies fade as stars die

5   Galaxies continue to separate

6   Galaxies at maximum separation

7   Galaxies start falling together

8   Galaxies start to merge

9   Big Crunch

△ A flat universe is the dividing line between an open universe and a closed one. In it, expansion will finally cease after an infinite amount of time has elapsed. A flat universe will be subject to proton decay and heat death in the same way as an open one. For all practical purposes, it is an open universe, and is distinguished in its own right merely because of the flatness problem.

△ A "closed" universe is one in which the force of gravity, created by the material it contains, is enough to halt the expansion of the Universe and draw it back together again. As the galaxies move closer, the Universe will heat up again, until an inevitable fireball, the Big Crunch, which is similar to but not the same as a reversal of the Big Bang. Some possible closed universes may be so long-lived that all the processes of an open universe, such as proton decay and heat death, can still take place before they collapse back in entirely.

EXTRATERRESTRIAL LIFE

HUMAN beings have always asked themselves whether the Earth is the only place in the Universe where life has developed. If the observations of giant molecular clouds result in the discovery of organic (carbon-bearing) molecules there, then any newly forming solar systems would be seeded with the chemicals necessary for life to develop. The presence of these building blocks is no guarantee, however, that life will emerge. The history of life on Earth seems so full of chance happenings that the likelihood of the same thing occurring on another world seems highly remote.

For life similar to our own to develop, the planet would have to resemble Earth in physical properties such as temperature, atmosphere and sunlight. This would only happen on a planet in orbit around a star similar to the Sun. The Sun is a G type star, but a cooler K type star might also suffice if the planet were a little closer to it. Hotter stars such as A and F types may also provide suitable homes for life-bearing planets, if they were to exist farther from the star than the Earth does from the Sun.

Detecting any such life-bearing planets from the Earth is extremely difficult. The Earth "leaks" radio broadcasts into space every day: perhaps other planets might do the same. Astronomers are actively listening

▽ NASA space probes Pioneer 10 and 11 carried this plaque because they were destined to leave the Solar System. They show the size of humans next to the probe. The starry diagram is a map which locates the Earth in relation to several nearby pulsars.

▷ The Arecibo radio telescope, situated in a natural crater in the mountains of Puerto Rico, is the largest in the world. It has a diameter of 305 meters and is used to scan different areas of the sky as they pass directly over the telescope.

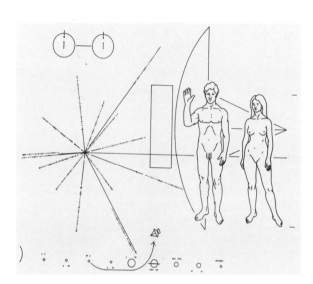

▽ In 1938 Orson Welles broadcast a dramatic radio adaptation of H. G. Wells' classic science fiction story *The War of the Worlds.* Set in contemporary America, it fooled many listeners into thinking that a Martian invasion was actually taking place. If that broadcast "leaked" into space, it would have arrived at nearby star systems during the years indicated. It is still moving outward, though the transmission is getting weaker all the time, and is probably unintelligible.

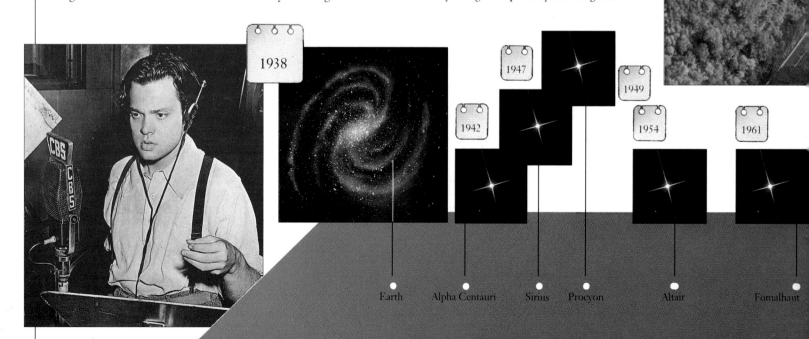

1938

1947

1949

1942

1954

1961

Earth   Alpha Centauri   Sirius   Procyon   Altair   Fomalhaut

▲ The contents of the "Arecibo message", broadcast into outer space in 1974 by a group of astronomers. The text, in binary code, included such diverse information as the binary numbers from 1 to 40; the atomic numbers of hydrogen, carbon, nitrogen, oxygen and phosphorus (five of the principal elements that make up life on Earth); chemical formulas and other details of DNA; a representation of a human being; the population of Earth; and a representation of the Solar System.

for these stray broadcasts. The most useful radio frequencies are in a section of the microwave spectrum known as the "waterhole". In this region of the spectrum, interstellar absorption of the radiation is at a minimum, as is atmospheric absorption. The name derives from two spectral lines that are contained within the region, one of hydrogen (chemical symbol H) and one of hydroxyl (a radical made of oxygen and hydrogen, OH). If the two are mentally placed together, H and OH give two hydrogens and an oxygen, H_2O, which is water. On this basis, this section of the microwave spectrum is known as the waterhole. In 30 years of listening, nothing has been discovered that seems beyond doubt to be a deliberate or accidental signal from another world, although one or two unexplained signals have been observed.

Astronomers on Earth have made some deliberate broadcasts as well as the "leaked" radio emissions. The first was aimed at globular cluster M13 and was sent in 1974 by astronomers using the 305-meter satellite dish at Arecibo, Puerto Rico. However, even though the message is traveling at the speed of light, it will take 24,000 years to reach M13. If the message were received, and a reply sent, it would take another 24,000 years to reach us. Many astronomers and engineers believe that in those intervening 48,000 years the human race will have developed the required technology to enable us to travel to the star and deliver the messages personally.

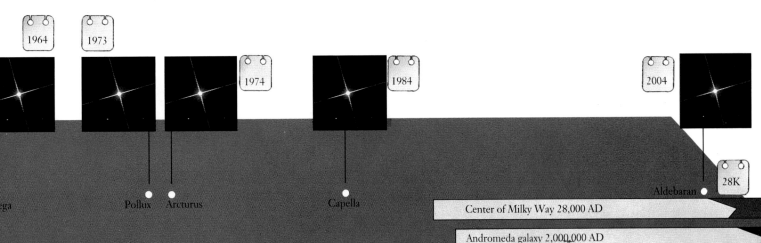

1964   1973   1974   1984   2004

28K

Vega   Pollux   Arcturus   Capella   Aldebaran

Center of Milky Way 28,000 AD

Andromeda galaxy 2,000,000 AD

2M

LIFE, MIND AND THE UNIVERSE

ALTHOUGH astronomers have a justifiable sense of accomplishment at understanding some of the processes that take place in the Universe, the more they learn about the Universe, the more questions seem to be thrown up. These questions are of a fundamental nature which science alone is not equipped to answer. Is humanity the only intelligence in the Universe? Is the Universe, and the human race, here by chance or as part of some grand design?

One question that has been asked by some modern astronomers is, why did the Universe turn out the way it did, when there were so many possible ways in which it could have been made? In the first few moments of the Big Bang, known as the Planck time, the laws of physics and the universal constants were in a state of flux. Only later did they become fixed into their familiar shapes and values. These laws of physics describe how the Universe behaves, and constants such as the speed of light provide the essential structures that we inhabit. If the Universe had a different constant charge on an electron, stars might never have been able to burn hydrogen; if the proportion by which matter outweighed antimatter in the first second of the Big Bang had been different, there could have been no matter at all, or so much that it would have collapsed long ago.

Even if such differences in constants had permitted the Universe to emerge, and even allowed life of sorts to evolve, the forms that such life might take would have been strikingly different. If the Planck constant, which governs interactions on the quantum scale, were much larger than its current value, even an object as large as a person might exhibit wave-particle duality and be capable of diffracting through a doorway in the same manner as electrons diffract through a tiny slit.

Philosophers may ask why a Universe so well adapted to our form of life ever came into being. Was it mere chance, or was the Universe made the way it is so that human beings could develop within it?

These questions are addressed by a highly controversial theory known as the anthropic cosmological principle. This states that the Universe exists as it does because, if it did not, we would not be here to observe it. One variation takes this further: the Universe exists in order to give the human race somewhere to live. Many object to this notion that the human race is somehow special, and point to the tenacious way in which life finds niches on the Earth in which to exist. This suggests that life will emerge wherever it has the slightest opportunity, an idea that can be applied to the Universe as a whole. Others argue that the Universe may not be unique, and there may have been earlier universes before the Big Bang; even that the laws of physics as we know them are the result of an evolutionary process through many previous cycles.

As time goes by, the Universe evolves more and more complex structures. At the simplest level are the fundamental particles, or quarks – the very first things that emerged after the Big Bang. At the most complex are intelligent beings and their conceptual constructs (including perhaps science itself, art and civilization). Such complex objects could not exist without the intervening emergence of structure through simple atoms, galaxies and stars, heavier elements, molecules, proteins, simple life forms and ever more organized life. Some argue that the evolution of intelligence is therefore as natural as the production of atoms and molecules. It may therefore be the purpose of intelligence to shape the Universe into a perpetual home for itself. In this way intelligence can give itself all the time it could need to explore and understand. Even if our own civilization fails, future civilizations may succeed in finding enough time to explore and understand the ultimate purpose, if one truly exists, of the Universe.

▷ **Marie Curie (1867–1934), pioneer in subatomic physics**

◁ **Galileo Galilei, one of the first classical physicists**

◼ At the Planck time, the only possible structures were quarks. As time went by, protons and neutrons formed, followed by electrons, which then combined to form atoms. These forged bonds to form simple molecules. As more complicated atoms were synthesized, much larger molecules such as organic ones containing carbon were formed. These molecules could then form living cells, which led to more complex, social life, such as bees. At the pinnacle of this progression is sentient, creative life such as humanity, exemplified by the composer Mozart.

1 Quarks

2 Nucleons

3 Atoms

4 Simple molecules

5 Macromolecules

6 Simple organisms

7 Social organisms

▽ **Wolfgang Amadeus Mozart, creative genius**

INTERSTELLAR TRAVEL

T HE ability to travel to distant stars would transform astronomy from an observational science into an experimental one. The problems involved in reaching the stars, however, are almost impossible to overstate. Scientists have sent probes to eight of the nine planets in the Solar System and have developed the fastest-moving artificial objects in the process. If those same probes were to be launched to the stars, they would take thousands of years to reach them. The distances to the stars are so huge that the light from the nearest star to the Sun – a triple star system known as Alpha Centauri A and B and Proxima Centauri – takes over four and a quarter years to reach the Earth. Because nothing in the Universe can travel faster than the speed of light, it seems that travel times on even the most advanced starships are liklely to be extremely long indeed.

Because of this, future astronauts may have to be placed in suspended animation. The metabolism of their bodies would be slowed down so that they fall unconscious. Computers would then monitor them and maintain their life so that their bodies age very slowly. Many years would pass as the starship

▷ **Even more exotic than interstellar travel is the possibility of traveling through time. According to one theory, it could be done in the vicinity of a rotating black hole. To do this, the time traveler would have to enter the ergosphere. This is the region in which the spacetime continuum is dragged around by the rotation of the black hole. If the spacecraft then were able to leave the black hole, without crossing the event horizon, some physicists think it might emerge at a point years in the past; or possibly even in a different Universe entirely.**

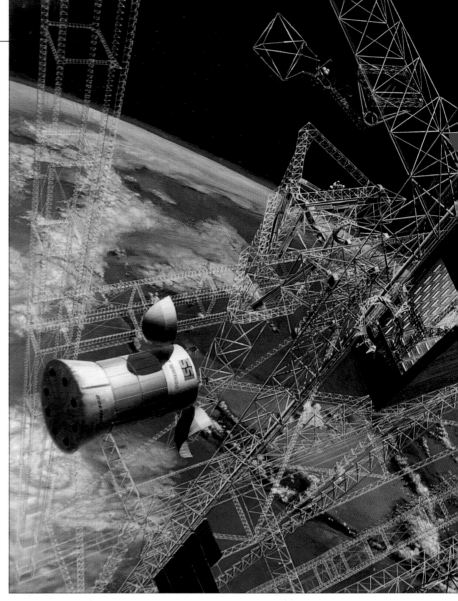

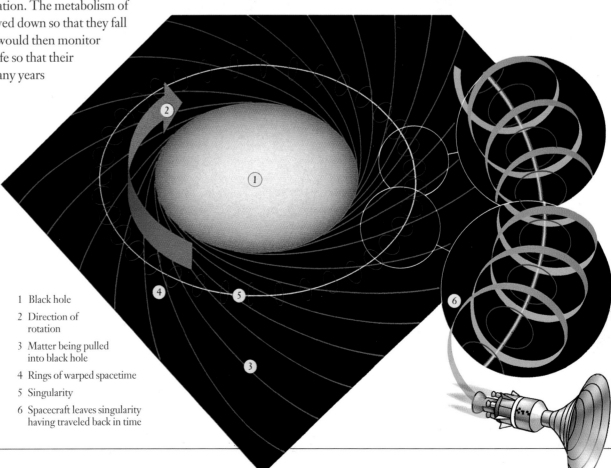

1 Black hole

2 Direction of rotation

3 Matter being pulled into black hole

4 Rings of warped spacetime

5 Singularity

6 Spacecraft leaves singularity having traveled back in time

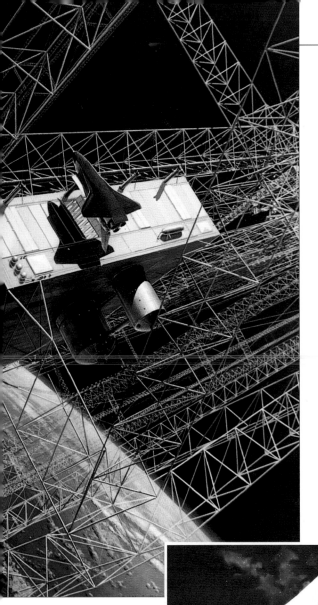

◁ This artist's impression shows a huge starship being constructed in orbit around the Earth – it would be far too large to build on a planet. The materials and the workforce would be ferried up and down by craft like a space shuttle. Plans for internationally financed space stations have proved difficult to bring to fruition; a starship, whose benefits would be very long-term indeed, would be even more problematic.

▷ The photon sail depends upon the principle of wave-particle duality. Because light rays can be particles (photons), they possess momentum. By bombarding the sail with photons from huge lasers, momentum could be transferred to the spacecraft.

▷ Carrying enough fuel for the journey is one problem of interstellar travel. The photon sail overcomes this one way, the ramscoop does it in another way. Seventy-five percent of the Universe is hydrogen, which is capable of nuclear fusion so, why not collect it en route? Conventional rockets accelerate the ramscoop to speed and then the "funnel" collects hydrogen, which is fused at the rear of the ship.

traveled to its destination under automatic controls. Upon arrival, the crew would be awakened. The mode of travel for this journey is known as the sleeper ship.

An alternative possibility is that the crew might live normal lives aboard what is known as a generation starship. As the original astronauts aged and died, their children would take over the running of the starship.

No propulsion system capable of reaching the stars has yet been built. Chemical rockets, such as those on the Space Shuttle, are not powerful enough to provide the thrust necessary for interstellar travel. Some scientists have proposed that nuclear bombs be detonated behind the starship to push it along. Another idea is to use powerful lasers and huge collecting mirrors. In the same way that sailing ships use sails to collect the wind, these starships would use mirrors to collect the photons of light. The radiation pressure of the photons would propel the starship. Finally, rockets powered by nuclear fusion might increase exponentially the speed at which they could travel through space: from $\frac{1}{20,000}$ of the speed of light to one-tenth the speed of light.

There are also highly exotic ideas which lie on the fringes of modern theoretical physics. If the Universe is made of many dimensions of which humans are only able to perceive three or four, perhaps a shortcut can be found through the other dimensions if they could be discovered. The mathematics of such "wormholes" are being calculated. If wormholes exist and can be accessed, and if they can be utilized for travel, then perhaps the entire Universe could become accessible.

The highly controversial possibility of traveling through time is also being investigated. Some astrophysicists believe that the warped spacetime continuum around a black hole is a potential time-traveling machine, but the practicalities of exploiting this – not to mention the dangers – preclude its foreseeable use to humans.

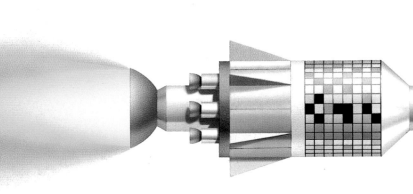

FACTFILE

Precise measurement is at the heart of all science, and the several standard systems have been in use in the present century in different societies. Today, the SI system of units is universally used by scientists, but other units are used in some parts of the world. The metric system, which was developed in France in the late 18th century, is in everyday use in many countries, as well as being used by scientists; but imperial units (based on the traditional British measurement standard, also known as the foot–pound–second system), and standard units (based on commonly used American standards) are still in common use.

Whereas the basic units of length, mass and time were originally defined arbitrarily, scientists have sought to establish definitions of these which can be related to measurable physical constants; thus length is now defined in terms of the speed of light, and time in terms of the vibrations of a crystal of a particular atom. Mass, however, still eludes such definition, and is based on a piece of platinum-iridium metal kept in Sèvres, near Paris.

□ METRIC PREFIXES

Very large and very small units are often written using powers of ten; in addition the following prefixes are also used with SI units. Examples include: milligram (mg), meaning one thousandth of a gram, kilogram (kg), meaning one thousand grams.

Name	Number	Factor	Prefix	Symbol
trillionth	0.000000000001	10^{-12}	pico-	p
billionth	0.000000001	10^{-9}	nano-	n
millionth	0.000001	10^{-6}	micro-	μ
thousandth	0.001	10^{-3}	milli-	m
hundredth	0.01	10^{-2}	centi-	c
tenth	0.1	10^{-1}	deci-	d
one	1.0	10^{0}	–	–
ten	10	10^{1}	deca-	da
hundred	100	10^{2}	hecto-	h
thousand	1000	10^{3}	kilo-	k
million	1,000,000	10^{6}	mega-	M
billion	1,000,000,000	10^{9}	giga-	G
trillion	1,000,000,000,000	10^{12}	tera-	T
quadrillion	1,000,000,000,000,000	10^{15}	exa-	E

□ CONVERSION FACTORS

Conversion of METRIC units to imperial (or standard) units

To convert:	to:	multiply by:
LENGTH		
millimeters	inches	0.03937
centimeters	inches	0.3937
meters	inches	39.37
meters	feet	3.2808
meters	yards	1.0936
kilometers	miles	0.6214
AREA		
square centimeters	square inches	0.1552
square meters	square feet	10.7636
square meters	square yards	1.196
square kilometers	square miles	0.3861
square kilometers	acres	247.1
hectares	acres	2.471
VOLUME		
cubic centimeters	cubic inches	0.061
cubic meters	cubic feet	35.315
cubic meters	cubic yards	1.308
cubic kilometers	cubic miles	0.2399
CAPACITY		
milliliters	fluid ounces	0.0351
milliliters	pints.	0.00176 (0.002114 for US pints)
liters	pints	1.760 (2.114 for US pints)
liters	gallons	0.2193 (0.2643 for US gallons)
WEIGHT		
grams	ounces	0.0352
grams	pounds	0.0022
kilograms	pounds	2.2046
tonnes	tons	0.9842 (1.1023 for US, or short, tons)
TEMPERATURE		
Celsius	fahrenheit	1.8, then add 32

Conversion of STANDARD (or imperial) units to metric units

To convert:	to:	multiply by:
LENGTH		
inches	millimeters	25.4
inches	centimeters	2.54
inches	meters	0.245
feet	meters	0.3048
yards	meters	0.9144
miles	kilometers	1.6094
AREA		
square inches	square centimeters	6.4516
square feet	square meters	0.0929
square yards	square meters	0.8316
square miles	square kilometers	2.5898
acres	hectares	0.4047
acres	square kilometers	0.00405
VOLUME		
cubic inches	cubic centimeters	16.3871
cubic feet	cubic meters	0.0283
cubic yards	cubic meters	0.7646
cubic miles	cubic kilometers	4.1678
CAPACITY		
fluid ounces	milliliters	28.5
pints	milliliters	568.0 (473.32 for US pints)
pints	liters	0.568 (0.4733 for US pints)
gallons	liters	4.55 (3.785 for US gallons)
WEIGHT		
ounces	grams	28.3495
pounds	grams	453.592
pounds	kilograms	0.4536
tons	tonnes	1.0161
TEMPERATURE		
fahrenheit	Celsius	subtract 32, then × 0.55556

□ SI UNITS

Now universally employed throughout the world of science and the legal standard in many countries, SI units (short for *Système International d'Unités*) were adopted by the General Conference on Weights and Measures in 1960. There are seven base units and two supplementary ones, which replaced those of the MKS (meter–kilogram–second) and CGS (centimeter–gram–second) systems that were used previously. There are also 18 derived units, and all SI units have an internationally agreed symbol.

None of the unit terms, even if named for a notable scientist, begins with a capital letter: thus, for example, the units of temperature and force are the kelvin and the newton (the abbreviations of some units are capitalized, however). Apart from the kilogram, which is an arbitrary standard based on a carefully preserved piece of metal, all the basic units are now defined in a manner that permits them to be measured conveniently in a laboratory.

Name	Symbol	Quantity	Standard
BASIC UNITS			
meter	m	length	The distance light travels in a vacuum in $1/299{,}792{,}458$ of a second
kilogram	kg	mass	The mass of the international prototype kilogram, a cylinder of platinum-iridium alloy, kept at Sèvres, France
second	s	time	The time taken for 9,192,631,770 resonance vibrations of an atom of cesium-133
kelvin	K	temperature	$1/273.16$ of the thermodynamic temperature of the triple point of water
ampere	A	electric current	The current that produces a force of 2×10^{-7} newtons per meter between two parallel conductors of infinite length and negligible cross section, placed one meter apart in a vacuum
mole	mol	amount of substance	The amount of a substance that contains as many atoms, molecules, ions or subatomic particles as 12 grams of carbon-12 has atoms
candela	cd	luminous intensity	The luminous intensity of a source that emits monochromatic light of a frequency 540×10^{-12} hertz and whose radiant intensity is $1/683$ watt per steradian in a given direction
SUPPLEMENTARY UNITS			
radian	rad	plane angle	The angle subtended at the center of a circle by an arc whose length is the radius of the circle
steradian	sr	solid angle	The solid angle subtended at the center of a sphere by a part of the surface whose area is equal to the square of the radius of the sphere

Name	Symbol	Quantity	Standard
DERIVED UNITS			
becquerel	Bq	radioactivity	The activity of a quantity of a radio-isotope in which 1 nucleus decays (on average) every second
coulomb	C	electric current	The quantity of electricity carried by a charge of 1 ampere flowing for 1 second
farad	F	electric capacitance	The capacitance that holds charge of 1 coulomb when it is charged by a potential difference of 1 volt
gray	Gy	absorbed dose	The dosage of ionizing radiation equal to 1 joule of energy per kilogram
henry	H	inductance	The mutual inductance in a closed circuit in which an electromotive force of 1 volt is produced by a current that varies at 1 ampere per second
hertz	Hz	frequency	The frequency of 1 cycle per second
joule	J	energy	The work done when a force of 1 newton moves its point of application 1 meter in its direction of application
lumen	lm	luminous flux	The amount of light emitted per unit solid angle by a source of 1 candela intensity
lux	lx	illuminance	The amount of light that illuminates 1 square meter with a flux of 1 lumen
newton	N	force	The force that gives a mass of 1 kilogram an acceleration of 1 meter per second per second
ohm	Ω	electric resistance	The resistance of a conductor across which a potential of 1 volt produces a current of 1 ampere
pascal	Pa	pressure	The pressure exerted when a force of 1 newton acts on an area of 1 square meter
siemens	S	electric conductance	The conductance of a material or circuit component that has a resistance of 1 ohm
sievert	Sv	dose	The radiation dosage equal to 1 joule equivalent of radiant energy per kilogram
tesla	T	magnetic flux density	The flux density (or density induction) of 1 weber of magnetic flux per square meter
volt	V	electric potential	The potential difference across a conductor in which a constant current of 1 ampere dissipates 1 watt of power
watt	W	power	The amount of power equal to a rate of energy transfer of (or rate of doing work at) 1 joule per second
weber	Wb	magnetic flux	The amount of magnetic flux that, decaying to zero in 1 second, induces an electromotive force of 1 volt in a circuit of one turn

An important concept in all forms of astronomy, the celestial sphere derives from ancient times, when the stars were thought to exist on a fixed sphere, surrounding the planets. All stars appear to be at the same distance away from Earth because human eyesight (or "vision") cannot detect the differences in their individual distances.

On Earth, we determine which stars are close by observing their motion in relation to other background objects. This exercise can be likened to using parallax with stars. Stellar parallax is so small that it is undetectable to the naked eye; hence we cannot judge the distances to stars.

Tiny portions of the celestial sphere are often referred to as the sky plane. Everything throughout space is projected onto this plane; stars, galaxies and nebulae appear within the same area of sky when, in fact, they are separated by thousands or even millions of light-years. Astronomers must sort out which objects are close and which are far away, so distance determination is

very important. Complicating our perception of the Universe is the fact that the celestial sphere is not a flat surface. The apparent distance between objects is misleading, and geometrical corrections for spherical surfaces must be made.

The celestial sphere is theoretically divided into areas known as constellations, and everything in them is referenced by coordinate systems. These are the equivalent of longitude and latitude on Earth. The simplest system defines everything by altitude and azimuth. The first equatorial system defines objects by hour angle and inclination. The best system uses coordinates known as right ascension and declination.

The apparent motion of the Sun on the celestial sphere is traced by a line, encircling the Earth, which is called the ecliptic. Zenith is the point directly overhead. The Earth's Equator, projected onto the celestial sphere, is known as the celestial equator.

▽ The equatorial co-ordinate system uses right ascension and declination. The "altitude", or declination, of a celestial object is measured perpendicular to the celestial equator and always remains the same. Declination is measured in degrees: zero at the Equator, 90° at the North Pole, and -90° at the South Pole. The second ordinate, known as the right ascension, is always measured east of the first point of Aries (shown here by its constellation symbol). This is the position on the celestial sphere where the celestial equator crosses the ecliptic. Right ascension is measured in hours, minutes and seconds, the full circle comprising 24 hours (the length of time for the sphere to complete one apparent revolution).

THE CELESTIAL SPHERE

North celestial pole

Sun

Ecliptic

Summer solstice

Autumnal equinox

N

Earth

S

23.5°

Celestial equator

Vernal equinox ♈

Winter solstice

Prime meridian

South celestial pole

▽ The altazimuth system is the earliest system used to map the sky. The altitude is the position of the star above the observer's horizon (rather than in relation to the celestial equator). The zenith (or point directly over the observer's head), rather than the celestial pole, provides the other key element of this coordinate system. The meridian on which the star falls is then measured in terms of how far east it is from the north point on the horizon, measured in terms of hours and minutes. This is called its azimuth. As the Earth rotates, this system requires exact knowledge of the time, calculated in terms of the position of the Earth in relation to the stars (sidereal time), rather than the Sun (solar time).

CELESTIAL COORDINATES

Star

Declination

δ

α

Right ascension

♈

△ The celestial sphere is fundamental to astronomy. It is used to reference every object in the sky. The apparent motion of the Sun in the sky is along the ecliptic. The position of the Sun defines the seasons. The prime meridian passes through the poles and the vernal equinox (also the first point of Aries).

ALTAZIMUTH SYSTEM

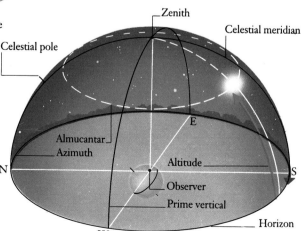

Zenith

Celestial meridian

Celestial pole

E

Almucantar

Azimuth

N

Altitude

S

Observer

Prime vertical

Horizon

W

☐ BRIGHTEST STARS

The ancient Greek astronomer Hipparchus (c. 130 BC) invented the first system to measure the brightness of stars. His system of six magnitudes lasted well into the 19th century, long after telescopes had begun to be used to view the stars. Stars are listed here in the order of their visible (apparent) magnitude – how bright they appear to an observer on Earth – rather than by their absolute magnitude, which is a measurement of their actual brightness.

· On the original visible magnitude scale, brightest stars were assigned a magnitude of 1 and the dimmest a magnitude of 6. By extension, 0 and negative values are even brighter than 1; high numbers are even dimmer than 6. Decimals graduate each stage (so that, for example, 1.3 is brighter than 1.8). Absolute magnitude is the brightness of a star as seen from an arbitrarily chosen distance of 10 parsecs (32.6 light-years) from the Earth.

Star	Name	Visible magnitude	Spec. class	Distance (light-yr)
Alpha Canis Majoris	Sirius	-1.5	AI V	8.6
Alpha Carinae	Canopus	-0.7	F0 I	1170
Alpha Centauri		-0.3	G2V	4.3
Alpha Boötis	Arcturus	0.0	K0 III	36
Alpha Lyrae	Vega	0.0	A0V	26
Alpha Aurigae	Capella	0.1	G8 III	42
Beta Orionis	Rigel	0.1	B8 I	910
Alpha Canis Minoris	Procyon	0.4	F5 IV	11
Alpha Eridani	Achernar	0.5	B5 IV	85
Alpha Orionis	Betelgeuse	0.5	M2 I	310
Beta Centauri	Hadar	0.6	B1 II	460
Alpha Aquilae	Altair	0.8	A7 IV	17
Alpha Tauri	Aldebaran	0.9	K5 III	68
Alpha Crucis	Acrux	0.9	B3 V	360
Alpha Scorpii	Antares	1.0	M1 I	330
Alpha Virginis	Spica	1.0	B1 V	260
Beta Geminorum	Pollux	1.1	K0 III	36
Alpha Piscis Austrini	Fomalhaut	1.2	A3 V	22
Alpha Cygni	Deneb	1.3	A2 I	1830
Beta Crucis	Mimosa	1.3	B0 III	420
Alpha Leonis	Regulus	1.4	B7 V	85
Epsilon Canis Majoris	Adhara	1.5	B2 II	490
Alpha Geminorum	Castor	1.6	A1 V	46
Gamma Crucis		1.6	M3 III	88
Gamma Orionis	Bellatrix	1.6	B2 III	360
Lambda Scorpii	Shaula	1.6	B2 IV	270

Star	Name	Visible magnitude	Spec. class	Distance
Alpha Gruis		1.7	B5 V	68
Epsilon Orionis	Alnilam	1.7	B0 I	1210
Beta Tauri	Elnath	1.7	B7 III	130
Beta Carinae		1.7	A0 III	85
Alpha Persei	Mirfak	1.8	F5 I	620
Zeta Orionis	Alnitak	1.8	O9 I	1110
Alpha Ursae Majoris	Dubhe	1.8	KO III	75
Epsilon Ursae Majoris	Alioth	1.8	A0p	62
Theta Scorpii		1.9	F0 I	910
Alpha Trianguli	Australe	1.9	K2 III	55
Beta Aurigae		1.9	A2 IV	72
Eta Ursae Majoris	Alkaid	1.9	B3 V	110
Gamma Geminorum		1.9	A0 IV	85
Epsilon Sagittarii	Kaus Australis	1.9	B9 IV	85
Delta Canis Majoris		1.9	F8 I	3070
Epsilon Carinae		1.9	K0 II	200
Alpha Pavonis	Peacock	1.9	B3 IV	230
Gamma Leonis	Algieba	1.9	K0 III	100
Beta Canis Majoris	Mirzam	2.0	B1 II	720
Alpha Hydrae	Alphard	2.0	K3	85
Alpha Ursae Minoris	Polaris	2.0	F8 I	680
Beta Ceti		2.0	K0 III	68
Delta Velorum		2.0	A0 V	68
Sigma Sagittarii	Nunki	2.0	B3 IV	210
Alpha Arietis	Hamal	2.0	K2 III	85

☐ NEAREST STARS

Simply because it is so close to Earth, the Sun – by no means an exceptional star, either in its size or its brightness – dominates our sky. Like most of the other stars closest to Earth, it is a main sequence star (type G2); one star on the list, Procyon B, is a white dwarf. By contrast, most of the stars on the brightest star list are giants. Statistically, this is important because it provides good evidence that smaller stars populate the galaxy in greater numbers than larger stars. If all kinds of stars existed in equal numbers, then giants should populate this list as well.

The table also lists each star's spectral class, which provides information about the temperature of the star. There are eight spectral classes for main sequence stars – O, B, A, F, G, K, M, and S. Of these, O represents the hottest type of star and S is the coolest.

Name	Distance (light-years)	Spec. class	Visible magnitude
Sun		G2 V	-26.7
Alpha Centauri	4.3	G2 V	-0.3
Barnard's Star	5.9	M5 V	9.5
Wolf 359	7.6	M8e	13.5
Lalande 21185	8.1	M2 V	7.5
Lutyen 726-8A	8.4	M6e	12.5
Sirius	8.6	A1 V	-1.5
Ross 154	9.4	M5e	10.6
Ross 248	10.3	M6e	12.3
Epsilon Eridani	10.7	K2 V	3.7
Luyten 789-6	10.8	M7e	12.2
Ross 128	10.8	M5	11.1
61 Cygni A	11.2	K5 V	5.2
61 Cygni B	11.2	K7 V	6.0

Name	Distance (light-years)	Spec. class	Visible magnitude
Epsilon Indi	11.2	K5 V	4.7
Procyon A	11.4	F5 IV-V	0.4
Procyon B	11.4	WD	10.7
Sigma 2398 A	11.5	M3.5 V	8.9
Sigma 2398 B	11.5	M4 V	9.7
Groombridge 34 A	11.6	M1 V	8.1
Groombridge 34 B	11.6	M6 V	11.0
Lacaille 9352	11.7	M2 V	7.4
Tau Ceti	11.9	G8 V	3.5
BD + 5° 1668	12.2	M5	9.8
L725-32 (YZ Ceti)	12.4	M5e	11.6
Lacaille 8760	12.5	M1	6.7
Kapteyn's star	12.7	M0 V	8.8
Kruger 60 A	12.8	M4	9.8

The northern sky in summer is dominated by the Summer Triangle. This consists of three stars: Deneb in the constellation Cygnus, Vega in Lyra and Altair in Aquila. The Summer Triangle is not a true constellation, but the phrase was coined by the British astronomer and broadcaster Patrick Moore to describe the most obvious shape in the summer sky. It is a very useful place to begin when attempting to find the other, fainter constellations.

Deneb itself is the brightest star in the constellation Cygnus (the Swan). To the north-east is the constellation Ursa Major (commonly called the Big Dipper or the Plow) – its six prominent second-magnitude stars are hard to miss. Two of its end stars (Dubhe and Merak) are sometimes called the Pointers, because an

imaginary line drawn through them leads directly to Polaris, the Pole Star and a sure indicator of the direction north. Vega is one of the brightest stars in the northern sky; like Arcturus to the east, in the constellation Boötes, it has a visible magnitude of 0. Altair, the third star in the Summer Triangle, is nearly as bright with a magnitude of 0.8.

The summer sky contains many deep sky objects, such as the Ring Nebula, a planetary nebula in Lyra; M13, a globular cluster of stars in Hercules; and the Veil Nebula, which is a supernova remnant in Cygnus.

Also visible in the northern sky in the summer months is the constellation Virgo, which contains a spectacular cluster of galaxies.

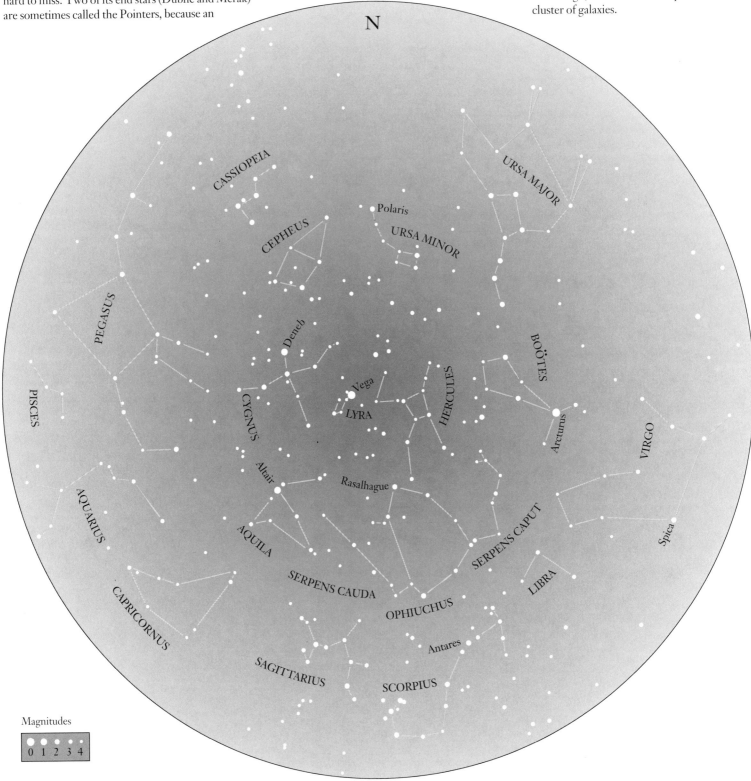

Magnitudes

0 1 2 3 4

In winter, the northern sky is dominated by the magnificent constellation of Orion, the Hunter. One of the most obvious constellations in the entire sky, it contains two zero-magnitude stars, Betelgeuse and Rigel. Orion's belt points toward Sirius, the brightest star in the sky. As well as bright stars, Orion also contains a nebula, visible to the naked eye as the second star in Orion's sword. Continued observation shows that it is, in fact, a fuzzy pink patch of glowing hydrogen. As bright as Rigel (magnitude 0.1) is Capella in the constellation Auriga (the Charioteer), which lies near the edge of the Milky Way between Orion and Ursa Major. Auriga is linked to Taurus (the Bull), which has its own 0.9 magnitude star, Aldebaran.

The winter sky is also ringed by the brightest of the zodiacal constellations: Taurus, Gemini and Leo. Gemini's brightest stars, the heavenly twins Castor and Pollux, have magnitudes of 1.6 and 1.1 respectively. Leo (the Lion) is a large constellation west of Gemini. Regulus, its most prominent star, is intermediate in brightness between Castor and Pollux. Ursa Major, although also visible in the summer, is now better placed for observation. The constellations relating to the myth of Perseus, of which the W-shaped Cassiopeia is the most easily detectable, are also visible to the east of the Pole Star. It contains a powerful radio and X ray source, the remnants of a supernova explosion that occurred in 1572.

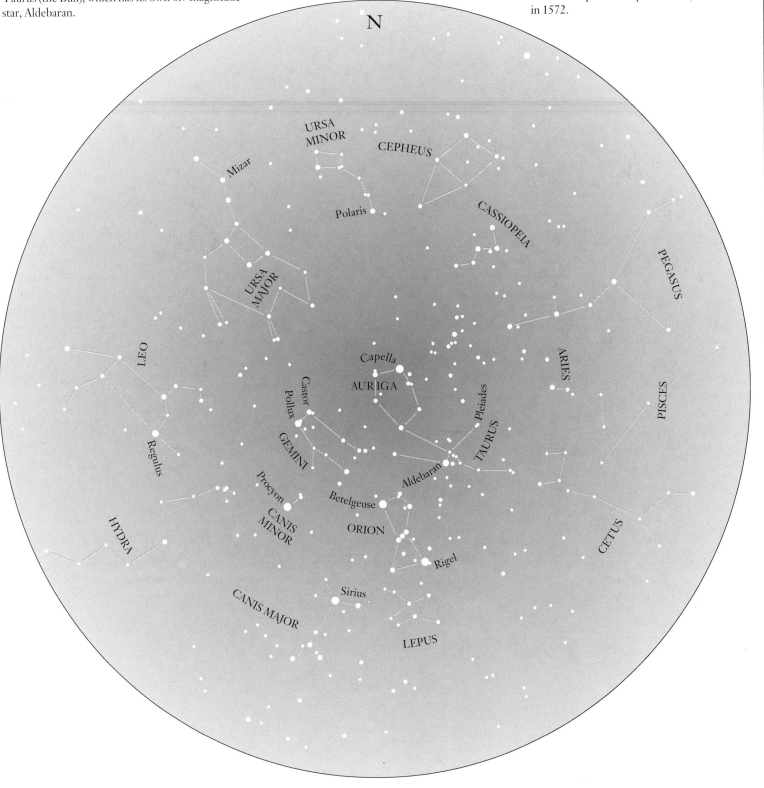

Whereas the constellation Orion is seen low in the winter sky of the Northern Hemisphere, in the Southern Hemisphere during summer it is almost directly overhead. Sirius in Canis Major shines with brilliant intensity and is free from the twinkling caused by its low altitude in the north.

The Milky Way, to which the Earth's Solar System belongs, winds its way through Centaurus, Crux, Vela, beyond Orion and through Auriga. It is always well placed for observation in the southern sky. Two satellite galaxies of the Milky Way are visible to the naked eye in the southern sky. The Large Magellanic Cloud can be seen about halfway between Carina and Hydrus, while the Small Magellanic Cloud lies within Tucana, near the boundaries of Hydrus.

The southern sky has some of the brightest stars in the sky. In addition to those that can sometimes also be seen from the north, they include Canopus (in the constellation Carina, the Keel) and Alpha Centauri (in Centaurus, the Centaur); their respective magnitudes of –0.7 and –0.3 make them, after Sirius, the second and third brightest stars.

The constellation Crux (the Cross) lies between Carina and Centaurus and contains the 0.9 magnitude star Acrux. Another bright southern star, Achernar (magnitude 0.5), lies to the east of Carina in the otherwise inconspicuous constellation Eridanus. The star itself gets its name from Arabic words that mean "end of the river", indicating its position in the constellation.

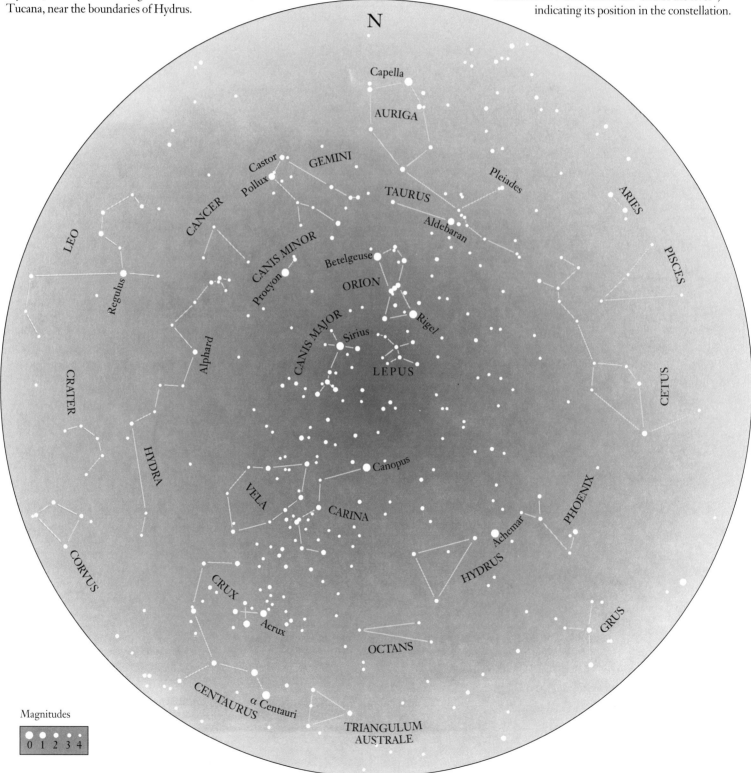

Magnitudes

0 1 2 3 4

The southern sky in winter contains one of the finest sights in the whole sky. It is during the winter months that the galactic center is virtually overhead; it is located in the constellation Sagittarius. The Milky Way strctches up and over the sky, like the handle on a shopping basket, from Carina in the south to Cygnus in the north. As it approaches Sagittarius, it widens out into an elliptical star cloud, which is our view of the Galaxy's central bulge.

Observers with binoculars or telescopes can see the vast number of globular clusters which surround the center of our Galaxy. The Northern Hemisphere's Summer Triangle is transformed into the Southern Hemisphere's Winter Triangle, with Deneb low in the north, and Vega and Altair slightly higher.

The Zodiacal constellations cut the Milky Way almost at right angles, starting with Virgo (the Virgin) in the east, then Libra (the Scales), Scorpius (the Scorpion) and Sagittarius (the Archer), ending with Aquarius (the Water Carrier) in the west. Some very bright stars are also visible in the winter sky, including the zero-magnitude Arcturus in Boötes (the Bear Driver); Spica, magnitude 1.0, in Virgo; and Formalhaut, magnitude 1.2, in the faint constellation of Piscis Austrinus (the Southern Fish). The south celestial pole is more difficult to locate than the north pole. It lies about half way along an imaginary line drawn from the upright part of the cross formed by Crux to the bright star Achernar. It is south of the Triangulum Australe (Southern Triangle).

Patterns of the stars have long been traced to illustrate myths or to mark the passage of the seasons. The maps presented here are portions of the celestial sphere containing some of the most interesting constellations. The area of sky designated to the constellation is marked with a line of white dashes. Right ascension lines are straight radial lines which are marked in hours; declination lines cross at right angles and are marked in degrees. The surrounding constellations are named, although the stars are not joined into the familiar patterns. Nebulas and galaxies are also marked on the maps. Stars in white are visible with binoculars; stars in black require more sophisticated equipment to be seen.

Magnitudes

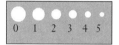

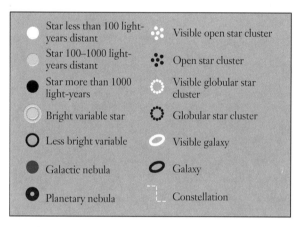

URSA MAJOR

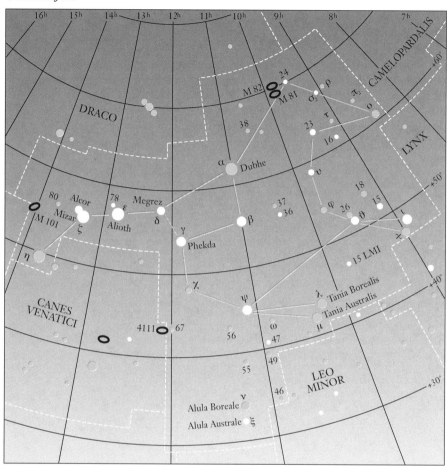

ORION

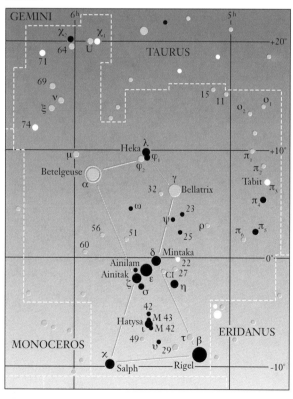

CENTAURUS

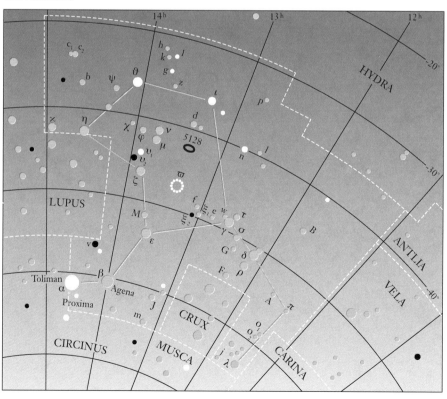

CASSIOPEIA

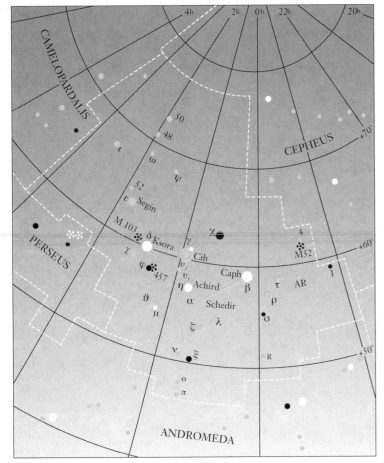

CYGNUS

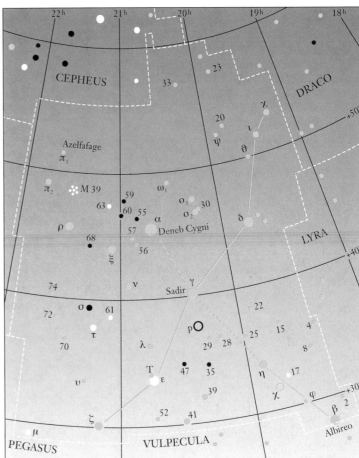

SAGITTARIUS

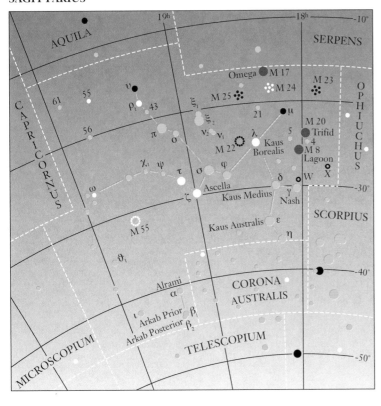

ANDROMEDA

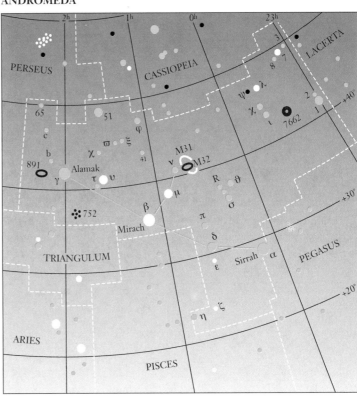

Forty-eight separate constellations were listed by Ptolemy, the Greek librarian at Alexandria, in the second century AD. In Europe, the study of constellations reached a peak around 1600, when the stellar cartographers, known as uranographers, produced fabulous works of art depicting the characters and creatures of the constellations. So enthusiastic were these uranographers that some left the stars themselves off the charts in favor of the artistic creatures.

Many of the early constellations were named after real or mythical animals, designated by their Latin names. There were birds (Crane, Crow, Eagle, Peacock, Phoenix, Swan), many animals (from Bears and Dogs to a Unicorn and Water Snake) and characters from mythology (such as Andromeda, Hercules and Perseus). The twelve constellations that lie along the Zodiac – an imaginary band that extends 8° each side of the ecliptic and in which the Moon and planets appear to move – took on particular significance, especially after the advent of astrology.

Around this time, other constellations were added to fill in the gaps left by the ancients. In those days, however, constellations were thought to be the actual patterns of the stars. More modern and pragmatic names were employed, such as Furnace, Microscope, Telescope and Triangle. Later, where large regions of sky were devoid of bright stars, new constellations were drawn using the faint stars. Constellations thus became areas rather than simply patterns of stars. As the study of stars progressed, there was a clear need for a system for referencing the location of faint stars and other objects. By international agreement in 1934, the celestial sphere was divided into 88 areas containing the familiar stellar patterns. They vary greatly in size, from Crux, the smallest, to the sprawling constellation of Hydra.

Latin name	Abbreviation	English name	Area
Andromeda	And	Andromeda	722
Antlia	Ant	Air Pump	239
Apus	Aps	Bird of Paradise	206
Aquarius	Aqs	Water Carrier	980
Aquila	Aql	Eagle	652
Ara	Ara	Altar	237
Aries	Ari	Ram	441
Auriga	Aur	Charioteer	657
Boötes	Boo	Bear Driver	907
Caelum	Cae	Chisel	125
Camelopardalis	Cam	Giraffe	757
Cancer	Cnc	Crab	506
Canes Venatici	CVn	Hunting Dogs	465
Canis Major	CMa	Big Dog	380
Canis Minor	CMi	Little Dog	183
Capricornus	Cap	Sea Goat	414
Carina	Car	Keel	494
Cassiopeia	Cas	Cassiopeia	598
Centaurus	Cen	Centaur	1060
Cepheus	Cep	Cepheus	588
Cetus	Cet	Whale	1231
Chamaeleon	Cha	Chamaeleon	132
Circinus	Cir	Compasses	93
Columba	Col	Dove	270
Coma Berenices	Com	Berenice's Hair	386
Corona Australis	CrA	Southern Crown	128
Corona Borealis	CrB	Northern Crown	179
Corvus	Crv	Crow	184
Crater	Crt	Cup	282
Crux	Cru	Cross	68
Cygnis	Cyg	Swan	804
Delphinus	Del	Dolphin	189
Dorado	Dor	Goldfish	179
Draco	Dra	Dragon	1083
Equuleus	Equ	Foal	72
Eridanus	Eri	River Eridanus	1138
Formax	For	Furnace	398
Gemini	Gem	Twins	514
Grus	Gru	Crane	366
Hercules	Her	Hercules	1225
Horologium	Hor	Clock	249
Hydra	Hya	Sea Monster	1303
Hydrus	Hyi	Water Snake	243
Indus	Ind	Indian	294

Latin name	Abbreviation	English name	Area
Lacerta	Lac	Lizard	201
Leo	Leo	Lion	947
Leo Minor	LMi	Little Lion	232
Lepus	Lep	Hare	290
Libra	Lib	Scales	538
Lupus	Lup	Wolf	334
Lynx	Lyn	Lynx	545
Lyra	Lyr	Lyre	286
Mensa	Men	Table	153
Microscopium	Mic	Microscope	210
Monoceros	Mon	Unicorn	482
Musca	Mus	Fly	138
Norma	Nor	Rule	165
Octans	Oct	Octant	291
Ophiuchus	Oph	Serpent Bearer	948
Orion	Ori	Hunter	594
Pavo	Pav	Peacock	378
Pegasus	Peg	Pegasus	1121
Perseus	Per	Perseus	615
Phoenix	Pho	Phoenix	469
Pictor	Pic	Easel	247
Pisces	Psc	Fishes	889
Piscis Australis	PsA	Southern Fish	245
Puppis	Pup	Poop	673
Pyxis	Pyx	Compass Box	221
Reticulum	Ret	Net	114
Sagitta	Sge	Arrow	80
Sagittarius	Sgr	Archer	867
Scorpius	Sco	Scorpion	497
Sculptor	Scl	Sculptor	475
Scutum	Sct	Shield	109
Serpens	Ser	Serpent	637
Sextans	Sex	Sextant	314
Taurus	Tau	Bull	797
Telescopium	Tel	Telescope	252
Triangulum	Tri	Triangle	132
Triangulum Australe	TrA	Southern Triangle	110
Tucana	Tuc	Toucan	295
Ursa Major	UMa	Great Bear	1280
Ursa Minor	UMi	Little Bear	256
Vela	Vel	Sails	500
Virgo	Vir	Virgin	1294
Volans	Vol	Flying Fish	141
Vulpecula	Vul	Fox	268

☐ HERTZSPRUNG–RUSSELL DIAGRAM

Every star in the night sky can be located on the Hertzsprung–Russell (H–R) diagram if the values for the axes of the graph are chosen carefully enough. The different regions cover the stages in stellar evolution. As stars age and evolve between these different stages, they move on the H-R diagram. If the movement is plotted between the regions, then an evolutionary track will be traced across the diagram, based upon the mass of the star. Higher-mass stars evolve more quickly than lower-mass stars, although nothing on the tracks indicates how much time is spent in each of the stages of evolution.

The Hertzsprung–Russell diagram has many different uses. If all the stars in the same cluster are plotted on the diagram, it will show the average age of the cluster. If the different types of variable star are plotted, the resulting diagram shows that different forms of variable star populate different regions on the diagram. For instance, in their late stages, all red giant stars become unstable and pulsate. If all types of variables are plotted, a strip on the diagram is shown to be defined. This has become known as the instability strip.

Stellar evolution

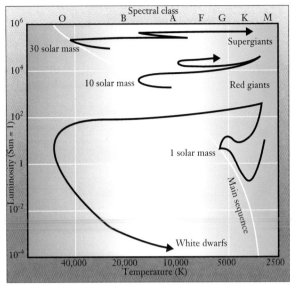

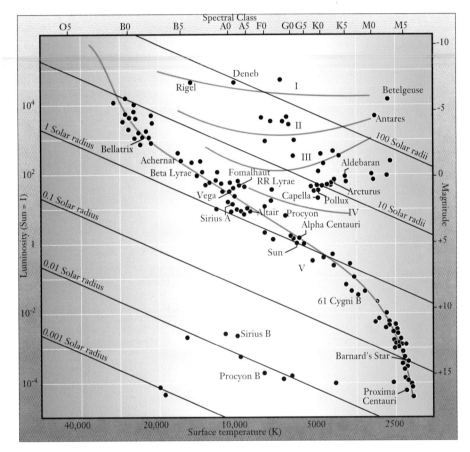

◁ **The Hertzsprung –Russell diagram** plots the temperature (or spectral class) and luminosity (or magnitude) of stars. It also provides information about their size, as the larger stars have higher luminosity and tend to be cooler, whereas small ones have low luminosity but are hotter. The diagram contains distinct areas: the main sequence on which most "middle-aged" stars are found; the regions of giants, toward which stars move when they leave the main sequence; and of white dwarfs at the bottom.

△ **Many aspects of the** history of stars can be plotted on the H–R diagram. Here, the evolution of a star of equivalent mass to the Sun is shown, from its origin and entry onto the main sequence, to its red giant phase and eventual fate as a planetary nebula surrounding a white dwarf. In contrast, the key stages of development of stars of ten and 30 solar masses are shown. After their red giant and supergiant phases, these high-mass stars will become supernovas, which cannot be plotted.

☐ FURTHER READING

Abell, George O., Morrison, David and Wolff, Sidney C. *Exploration of the Universe* (Saunders College Publishing, 1987)

Brandt, John C. and Chapman, Robert D. *Rendezvous in Space* (W.H. Freeman and Co., Oxford, 1992)

Chown, Marcus *Afterglow of Creation* (Arrow Books, London, 1993)

Cohen, Nathan *Gravity's Lens* (John Wiley and Sons, Chichester, 1988)

Disney, Michael *The Hidden Universe* (J.M. Dent, London, 1984)

Fabian, A.C. (ed) *Origins* (Cambridge University Press, 1988)

Ferris, Timothy *Galaxies* (Bantam Press, London 1988)

Gowstad, John *Astronomy: The Cosmic Perspective* (John Wiley and Sons, Chichester, 1990)

Gribbin, John *Blinded by the Light* (Bantam Press, London, 1991)

Gribbin, John *The Omega Point* (Heinemann, Oxford and London, 1987)

Hawking, Stephen *A Brief History of Time* (Bantam Press, London, 1988)

Henbest, Nigel (ed) *Observing the Universe* (Basil Blackwell and New Scientist, Oxford, 1984)

Kauffman, William J. III *Universe* (W.H. Freeman and Co., Oxford, 1985)

Kippenhahn, R. *100 Billion Suns* (Weidenfeld and Nicholson, London, 1983)

Kitchin, C.R. *Astrophysical Techniques* (Adam Hilger, Bristol, 1991)

Kitchin, C.R. *Journeys to the Ends of the Universe* (Adam Hilger, Bristol, 1990)

Kitchin, C.R. *Stars, Nebula and the Interstellar Medium* (Adam Hilger, 1987)

Levy, David *The Sky: A User's Guide* (Cambridge University Press, 1991)

Littman, Mark *Planets Beyond* (John Wiley and Sons, Chichester, 1988)

Nicolson, Iain and Moore, Patrick *The Universe* (Collins, London, 1985)

Osterbrock, Donald G. (ed) *Stars & Galaxies: Citizens of the Universe* (W.H. Freeman, Oxford, 1990)

Parker, Barry *Invisible Matter and the Fate of the Universe* (Plenum Press, London, 1989)

Rowan-Robinson, Michael *Cosmic Landscape* (Oxford University Press, 1979)

Schaaf, Fred *Wonders of the Sky* (Dover Publications, 1983)

Tucker, Wallace and Riccardo Giacconi *The X-Ray Universe* (Harvard University Press, 1985)

Weinberg, S. *The First Three Minutes* (Andre Deutsch, London, 1977)

Zukav, Gary *The Dancing Wu Li Masters* (Flamingo, 1979)

ACKNOWLEDGMENTS

Picture credits

1 SPL/AATB/Royal Observatory, Edinburgh **2** SPL/R Ressmeyer, Starlight **3** SPL/Phillipe Plailly **6** Kitt Peak Observatory **7** SPL/ NOAO **48–49** SPL **50–51** SPL/ Roger Ressmeyer, Starlight **52–53** SPL/David Parker **53** SPL/Philippe Plailly **56** SPL/Roger Ressmeyer, Starlight **56–57** SPL/Roger Ressmeyer, Starlight **59tr** TIB/Alan Becker **62** SPL/C Powell, P Fowler & D Perkins **64** Brinbo Books **66–67** SPL/US Naval Observatory **68t** Allsport/ Bob Martin **68b** SPL/John Reader **68–69** SPL/Roger Ressmeyer **69** The Photo Source/VCL **70–71** SPL/ Irvine-Michigan – Brookhaven proton decay experiment **71** SPL/David Parker **74–75** SPL/MaGrath Photography **76–77** Dr Nigel Metcalfe, University of Durham/ Royal Greenwich Observatory **77** AOL **82–83** SPL/Royal Observatory, Edinburgh **84–85** SPL **85** AOL/ Hatfield Polytechnic Observatory **86–87** SPL **88–89** SPL **90–91** SPL/NOAO **91** SPL/ Max Planck Institut für Radioastronomie **92** SPL/Roger Ressmeyer, Starlight **92–3** SPL/ NOAO **94** SPL **96** AOL/NASA **97t** SPL/ Hale Observatories **97c** SPL/ NASA **97b** SPL/B Cooper & D Parker **98–99** SPL **99** SPL **100–101** SPL/NOAO **102–103** SPL **104–105** NASA **106–107** AOL/Kitt Peak Observatory **108–109** SPL/David Nunuk **110** AOL/ Lick Observatory, California **111** SPL/ Royal Observatory, Edinburgh **112bl** R W Forrest l994 **112bc** R W Forrest l994 **112br** R W Forrest l994 **115** F Paresce, R Jedrezejewski, NASA (StScI) **116–117** SPL **118–119** SPL/NASA/ Space Telescope Science Institute **119t** OSF/Sean Morris **119c** OSF/Sean Morris **119b** OSF/Sean Morris **121l** SPL/NASA **121r** SPL/ NASA **122** SPL/NASA **122–123** SPL/AATB/ Royal Observatory, Edinburgh **123** SPL/ Hale Observatory **124** SPL/Kim Gordon **126–127** SPL/Royal Observatory, Edinburgh **127t** SPL/Dr S Goll & Dr J Fielden **130–131** SPL/NASA/Space Telescope Science Institute **131l** AOL/ Universität Tübingen **131c** AOL/ Universität Tübingen **131r** AOL/ Universität Tübingen **132–133** SPL **134–135** Dr Nigel Metcalfe, University of Durham/ Royal Greenwich Observatory **138bl** Popperfoto **138–139t** SPL **138–139b** Galaxy Picture Library **140bl** SPL **140br** SPL/National Library of Medicine **141** AOL **142–143** AOL/Boeing Aerospace Company **143** NASA

Abbreviations

b = bottom, **t** = top
l = left, **c** = center, **r** = right

AOL   Andromeda Oxford Limited, Abingdon, UK
SPL   Science Photo Library, London, UK

Artists

Mike Badrock, Rob and Rhoda Burns, John Davies, Hugh Dixon, Bill Donohoe, Sandra Doyle, John Francis, Shami Ghale, Mick Gillah, Ron Hayward, Jim Hayward, Trevor Hill/Vennor Art, Joshua Associates, Frank Kennard, Pavel Kostell, Ruth Lindsey, Mike Lister, Jim Robins, Colin Rose, Colin Salmon, Leslie D. Smith, Ed Stewart, Tony Townsend, Halli Verinder, Peter Visscher

Editorial assistance

Peter Lafferty, Ray Loughlin, Iain Nicolson, Lin Thomas

Index

Ann Barrett

Origination by

HBM Print Ltd, Singapore; ASA Litho, UK